SP–4027

I0493582

ASTRONAUTICS AND AERONAUTICS, 1986–1990

A Chronology

by Ihor Y. Gawdiak, Ramón J. Miró, and Sam Stueland

THE NASA HISTORY SERIES

National Aeronautics and Space Administration
NASA History Office
Office of Policy and Plans
Washington, DC   1997

Preface

This chronology of events in aeronautics, aviation, space science, and space exploration was prepared by the Federal Research Division of the Library of Congress for the History Division of the National Aeronautics and Space Administration (NASA). It covers the years 1986–1990 and continues the series of annual chronologies published by NASA.

The present volume returns to the format used in the *Astronautics and Aeronautics, 1979–1984: A Chronology* volume. It also integrates in a single table the information presented in two or three previous publications.

Ihor Y. Gawdiak
Ramón J. Miró
Sam Stueland

May 1997

Contents

January

January 2: A Reuters report said that Arabsat-1, the Arab world's first communication satellite, built for the Arab League by a French-led consortium and launched in February 1985 by a European Aryan rocket, failed to operate in orbit because of technical errors. The report also noted that Iraqi Director General Ali Mashat, the man in charge of the project, was fired last month by a committee of six Arab League communications ministers. (*C Trib*, Jan 2/86)

• Having examined NASA's 1986 schedule of launches, missions into space, and other space science events, U.S. media predicted that 1986 would "open an extensive new chapter in space exploration," and would rival the era of space exploration "that began with Galileo's first peak through a telescope 400 years ago." Predicting 1986 to be "a banner year," the *Christian Science Monitor* pictured an expanding role for the United States in space exploration because the year promised to put NASA on the cutting edge. Although the *Los Angeles Times* predicted "a year of spectaculars," it warned that no new starts for planetary missions beyond 1986 were on the horizon. The Times also noted that the projected $8 billion orbiting Space Station would likely take a big bite out of NASA funds and that increased military applications equated to less sharing of knowledge with the public. (*CSM*, Jan 2/86; *LA Times*, Jan 2/8)

January 3: The steering problem that halted the Shuttle Columbia's launch in December of 1985 was traced to a tiny electrical component that led ground computers astray. (*W Times*, Jan 3/86; *NY Times*, Jan 3/86)

• The Federal Aviation Administration has become increasingly concerned with the problem of inadequate airline maintenance. Because they maintain a 17-hour-per-day schedule and are involved in repeated takeoffs and landings, commercial jetliners are prone to numerous mechanical breakdowns. Although "modern jets are designed to fly despite even serious defects" and although U.S. carriers employ an army of airline mechanics repairing planes in between scheduled flights, several U.S. carriers were fined by the Federal Aviation Administration last summer for a "series of maintenance-related transgressions." The National Transportation Safety Board was considering whether a crash of a Midwest Express Airline jet in September 1985, in which 31 people were killed, was caused by poor maintenance. Some air-safety experts expressed fear that airline companies had become too cost conscious about safety; still, maintenance error ranked third behind pilot error and bad weather as a cause of fatal airline crashes. (*WSJ*, Jan 3/86)

January 6: NASA renamed two planetary missions scheduled for flight in 1988 and 1990. The mission to map the planet Venus, previously known as Venus Radar Mapper and scheduled for launch from the Space Shuttle in April 1988 and arrival at Venus in July 1989, was renamed Magellan. The Magellan spacecraft, attached to a Centaur-G upper stage, would map the entire surface of the planet Venus for the first time, using a synthetic-aperture radar instrument. The spacecraft would orbit the planet about once every three hours, coming as close as 250 kilometers from the surface.

The mission to map the planet Mars, previously known as the Mars Geoscience/Climatology Orbiter and scheduled for launch from the Space Shuttle in August 1990 and arrival at Mars in August 1999, was renamed Mars Observer. The Observer spacecraft, adapted from an existing production-line type of Earth-orbital spacecraft to reduce costs, would map the planet Mars to determine the global elemental and mineralogical character of its surface and to investigate the Martian climate, both present and past. The Magellan and Mars Observer projects were managed by the Jet Propulsion Laboratory in Pasadena, California, for NASA's Office of Space Science and Applications. (NASA Release 86-1)

• NASA announced that the launch of the Hubble Space Telescope and the launch of the first Earth Observation Mission have been switched. Under the new schedule, the Space Telescope would be launched on October 27, 1986, and the Earth Observation Mission would lift off on August 18, 1986. The change was made to provide additional time for delivery of the Space Telescope from the West Coast to Kennedy Space Center, Florida. The Space Telescope would be deployed into orbit from Space Shuttle Atlantis. It was expected to see 7 times farther and 10 times more clearly than any telescope on Earth. (NASA Release 86-2)

• *USA Today* reported that 40-year-old Karen Kofoed, a reporter for the *Camden Courier-Post,* was the first journalist to apply to be part of a crew aboard a Space Shuttle into space. As of January 6, only 20 of the 4000 reporters who requested applications had submitted the completed forms, which were due by January 15. The contest for the first journalist in space was administered for NASA by the Association of Schools of Journalism and Mass Communications at the University of South Carolina. (*USA Today,* Jan 6/86)

• A number of American space scientists expressed disappointment because no U.S. spacecraft would fly by Halley's Comet as it swings by Earth for the first time since 1910 and for the thirtieth time since 240 B.C. The scientists ascribed U.S. failure to have a mission to the comet to a "combination of over expectations, missed signals, poor planning, a streak of bad luck, and politics and infighting in the science community."

In the meantime, the European Space Agency's spacecraft *Giotto*, the Soviet Union's twin Vega spacecraft, and two Japanese spacecraft would each probe Halley's Comet. Particularly galling to American space scientists was the knowledge that none of these spacecraft could carry out their mission without U.S. help, as each was guided by the U.S. Deep Space Network to reach the comet. The most sophisticated mission would be by the spacecraft Giotto. After a 60,000 mile voyage through Halley's "dusty tail," it would pass "within 300 miles of the coma, which with the nucleus forms the comet's head." The twin Soviet spacecraft were launched in December 1984 into a Venus orbit, from which they landed probes on the planet and used that planet's gravity to fly toward Halley's comet. One craft was expected to stay about 6,000 miles from the nucleus, while the other would move closer. Japan's spacecraft would come only within 120,000 miles of Halley's comet. *(W Post,* Jan 6/86)

• For the third time in 19 days, launching of the Space Shuttle Columbia was again scrubbed. This time a malfunction with Columbia's computer control failed to close an oxygen valve. Engineers were successful in closing the valve under manual control, but the additional time it was open resulted in 1,500 gallons to 3,000 gallons of oxygen entering the fuel lines to the main engine, lowering temperatures to an unacceptable level. With 15 flights slated for 1886, another attempt was immediately planned for January 7.

The chief goal of the mission was to carry a $50 million communications satellite, Satcom, for RCA. When an RCA engineer announced that 8:47 p.m. was the cut off point for a launch that would later deliver the satellite at an optimum time, the mission was canceled. Scheduled to ride into space with NASA's astronauts and engineers was Florida's Democratic Congressman Bill Nelson, chairman of the House subcommittee on space science and applications that oversees NASA's budget. *(NY Times,* Jan 7/86; *W Post,* Jan 7/86; *P Inq,* Jan 7/86; *C Trib,* Jan 7/86; *W Times,* Jan 7/86; *B Sun,* Jan 7/86)

January 7: The scheduled launch of the Space Shuttle Columbia was again scrubbed—this time because of inclement weather at landing sights in Spain and the Sahara Desert. The weather situations in these distant areas were of concern to engineers at the Kennedy Space Center because the sites provided emergency landing in the event of an engine failure during the first few minutes of flight. This fifth delay, four in three weeks, raised questions concerning the NASA's ability to launch 15 scheduled Shuttle flights during 1986. *(W Post,* Jan 8/86; *NY Times,* Jan 8/86; *USA Today,* Jan 8/86; *P Inq,* Jan 8/86)

• The National Oceanic and Atmospheric Administration (NOAA), which operates the Nation's weather satellites, announced that its polar orbiting satellite, NOAA-8, was apparently lost. NOAA was informed by the Air Force that its tracking instruments detected "what appeared to be pieces of the satellite" in the vicinity of NOAA-8 orbit. *(NY Times,* Jan 7/86)

• Dr. William R. Graham, who was appointed Acting Administrator of NASA on January 3, 1986, stated in his first interview that he was the "acting administrator in every sense of the word" and that he felt very comfortable in his new position. He apparently made that statement to put to rest suggestions that the Agency would really be run by NASA Associate Administrator Philip E. Culberston while Graham learned the ropes. Graham was promoted to his new position, from the post of Deputy Administrator of NASA, when NASA Administrator James E. Beggs took a leave of absence to prepare his defense for a trial. Beggs had been charged with defrauding the Federal Government when he served as a top executive of General Dynamics Corporation.

NASA's acting administrator received his master's and doctoral degrees in engineering from Stanford University. He began his professional career at the Air Force Weapons Laboratory in New Mexico and then moved to the Rand Corporation in California. In 1971, he founded R&D Associates—a think tank in Marina Del Rey, California. He was a consultant to the Secretary of Defense and had served on an advisory committee on undersea warfare and human assisted strategic Air Force systems and on nuclear weapons panels. Prior to his appointment as Deputy Administrator of NASA in November 1985, Graham headed, for three years, President Reagan's advisory committee on arms control and disarmament.

The then 48-year-old native Texan expressed great optimism in his assessment of the future of NASA and of space exploration in general. He said that "the Agency in the coming decade would broaden mankind's understanding of Earth's environment, expand the reach of manned exploration, and extend scientific vision to the far edges of the universe." Stating that space exploration was limited only by "imagination," Graham promised to vigorously pursue President Reagan's mandate to make space more accessible to the ordinary citizen and free enterprise. (*W Times*, Jan 6/86; 7/86)

January 8: For the sixth time, launching of the Shuttle Challenger was scrubbed, this time because of a stuck fuel valve in one of the main engines. (*CSM*, Jan 9/86; *W Times*, Jan 9/86; *NY Times*, Jan 9/86; *P Inq*, Jan 9/86)

January 9: NASA reported that scientists observed Halley's comet from a University of Arizona telescope aboard NASA's Kuiper Airborne Observatory, a modified C-141 aircraft operated by NASA's Ames Research Center in Mountain View, California. It was the first direct confirmation of water in a comet, and the discovery lent new support to astronomers' widely held theory that comets are "dirty snowballs" composed primarily of frozen water. The theory was conceived by Dr. Michael Mumma, head of the Planetary Systems Branch, Goddard Space Flight Center, Greenbelt, Maryland, who, together with Dr. Harold Weaver, an associate research scientist at the Center for Astrophysical Sciences at Johns Hopkins University, developed a theoretical model leading to this discovery. (NASA Release 86-4; *W Post*, Jan 10/86; *B Sun*, Jan 10/86; *NY Times*, Jan 10/86)

January 10: The cause of the faulty engine valve that was responsible for Shuttle Columbia's sixth delay was discovered to be a 5-inch thermometer that was broken during fuel loading. After engineers removed the thermometer, they were hopeful for a launch later in the day, but feared that overcast weather might cause yet another delay. (*W Times*, Jan 10/86; *W Post*, Jan 10/86; *B Sun*, Jan 10/86)

January 12: Columbia, flight STS 61-C, was successfully launched from Cape Canaveral, Florida, after a Space Shuttle record of seven postponements. The crew completed their mission of launching RCA's $50 million satellite into orbit, for which NASA received $14.2 million from the corporation.

Ironically, after a 25-day delay in getting off the ground, the Space Shuttle was told to come down a day early. NASA engineers feared that bad weather might hamper the original landing date. Already behind schedule, NASA wanted extra time to prepare the Shuttle for the next flight in March and its mission of observing Halley's Comet. The needed orbit for studying the comet would leave little room for delays of this launch. Most of Columbia's experiments were completed, but a camera device for photographing Halley's comet failed to work because a light intensifier had been inadvertently left on during the ship's seven launch delays. It was not until January 18, after a two-day delay because of poor weather, that Columbia landed at Edwards Air Force Base in California. Engineers had hoped to land the craft at the Kennedy Space Center in Florida, from where it was scheduled to be launched on March 6. Therefore, an estimated five days were scheduled to be cut from its preparation time so that the critical date for observing Halley's Comet could be met. (*NY Times*, Jan 13/86; Jan 14/86; Jan 15/86; Jan 24/86; *USA Today*, Jan 13/86; Jan 15/86; *W Times*, Jan 13/86; *CSM*, Jan 13/86; Jan 21/86; *P Inq*, Jan 13/86; *B Sun*, Jan 13/86; *W Post*, Jan 13/86; Jan 15/86; Jan 16/86; *C Trib*, Jan 13/86)

January 14: NASA and SPACEHAB signed a memorandum of understanding (MOU) to establish cooperation for the latter's efforts to develop and market payload bay habitable modules, truncated metal cylinders that would be additions to the orbiter's crew department and connected by tunnel adapters. These modules, designed to increase the pressurized volume of the Space Shuttle, would serve as additional living and work space for the astronauts by providing an additional 1,000 cubic feet.

Another MOU, signed by NASA and the European Space Agency (ESA), was intended to give NASA access to Synthetic Aperture Radar data from the first European Remote Sensing Satellite for Government research purposes at the Fairbanks, Alaska, station that NASA was developing in connection with its Navy Remote Ocean Sensing Satellite System Scatterometer (NROSS). The data received from the ERS-1 satellite, engineers hoped, would enhance NASA-supported polar ice research, NROSS, the Topography Experiment for Ocean Circulation and Shuttle Imaging Radar-C. NASA was then to exchange its Scatterometer and radar imagery for other ERS-1 data of interest. The ERS-1 was scheduled for launch in 1989. (NASA Release 86-6; 86-7; *B Sun*, Jan 17/86)

January 16: The Voyager 2 Spacecraft discovered six new moons orbiting the planet Uranus, ranging in diameter from 20 to 30 miles. The moons were named 1986U1 in succession to 1986U6, denoting the year in which they were discovered, the planet they orbit and the order in which they were discovered. The first of these moons was found on January 8, 1986. (NASA Release 86-8; *W Post*, Jan 9/86; *CSM*, Jan 9/86; *NY Times*, Jan 9/86; *P Inq*, Jan 17/86; *B Sun*)

• Lasers on board weather satellites (LIDAR), said Wayne Baker of the Goddard Space Flight Center's Laboratory, promised more accurate forecasting. In the past, satellite data offered only marginal improvements over traditional methods and even deteriorations for some predictions. With the addition of a laser, however, scientists would have access to wind measurement throughout the atmosphere, keys to following the weather that were before limited to ground stations and weather balloons. (*CSM*, Jan 16/86)

January 21: The first Soviet Space Shuttle, expected to be launched in 1986, would likely look very similar to U.S. orbiters, said scientists. Photographs of a scale model Soviet space plane that splashed down in the Indian Ocean revealed a miniature American Shuttle. Plans for the plane were easy for scientists to obtain as the program was never top secret and sources noted that the Soviets had often copied Western aircraft in the past.

On a similar note, a report concerning the secretive launch from, presumably, the Soviet Union in the summer of 1985 concluded that an antisatellite weapon had gone up. Radar tracked debris that had orbited the Earth before burning up in the atmosphere, and researchers quickly deduced that the metallic scrap had come from a satellite on which the weapon was tested. The illegal launches were the first such unannounced flights by the Soviet Union since testing of orbital thermonuclear weapons. (*W Times*, Jan 21/86; *WSJ*, Jan 21/86)

January 23: Voyager 2 gathered a variety of technical information and sent back photographs of the planet Uranus, giving scientists their greatest knowledge yet of the distant planet. Among the discoveries were additional moons (bringing the total to 14), wind and clouds in the atmosphere, a brown, smog-like haze (probably caused by methane gas), and the existence of a magnetic field. (NASA

Release 85-165; *CSM,* Jan 22/86; Jan 23/86; Jan 24/86; *P Inq,* Jan 22/86; Jan 23/86; Jan 24/86; *W Post,* Jan 23/86; Jan 24/86; *USA Today,* Jan 22/86; *NY Times,* Jan 23/86; Jan 24/86; *B Sun,* Jan 24/86; *C Trib,* Jan 24/86)

January 26: For the second time in two days, launch of the Shuttle Challenger at Cape Canaveral, Florida, was postponed because of bad weather. Meanwhile, NASA began conducting tests at Edwards Air Force Base in California to determine the effect of rain on insulating tiles and readied a sophisticated instrument landing system, hoping to minimize the weather factor in launches and landings for future orbiter flights at Kennedy Space Center in Florida. (*W Times,* Jan 27/86; *USA Today,* Jan 27/86; *P Inq,* Jan 27/86; *C Trib,* Jan 27/86; *B Sun,* Jan 27/86)

January 28: Only 74 seconds after liftoff, the Space Shuttle Challenger, flight STS 51-L, exploded at about 10 miles above the Earth, killing all seven crew members. It was the worst accident in the history of space exploration and the first time American fatalities occurred during space flight. Slow motion film of the launch revealed a thin flame between the left booster rocket and the Shuttle's main fuel tank, causing engineers to speculate that hydrogen from a ruptured tank ignited upon contact with air heated by the rocket motors or directly by the motors themselves. It was hoped that investigation of the debris that fell into the Atlantic Ocean could lend a better understanding of the accident. All crew tended spaceflight was halted until a thorough investigation was completed, stunting the ambitious space program that promised to bring the richest lode of scientific data ever imagined. (*W Post,* Jan 29/86; *W Times,* Jan 29/86; *NY Times,* Jan 29/86; *B Sun,* Jan 29/86; *P Inq,* Jan 29/86; *C Trib,* Jan 29/86)

February

February 2: Initial investigation, and tape review of the Challenger accident, pointed to the right solid fuel rocket booster and showed a leak near one of its seams. An apparent rupture between two sections of the rocket emitted flames that soon found their way to the external fuel tank, igniting the volatile contents.

Meanwhile, salvage operations recovered two large objects believed to be part of the Shuttle from the ocean floor, but most of the craft was still unaccounted for. NASA planned to bring in a recovery vessel equipped with a remote-controlled robot and cameras. (*WSJ,* Feb 3/86; *NY Times,* Feb 3/86; *W Post,* Feb 3/86)

February 3: President Ronald Reagan issued an executive order through his press secretary that established a commission to investigate the Challenger accident. According to the order, the *ad hoc* commission would review the circumstances surrounding the accident, develop recommendations for corrective or other action based on the findings, and submit a final report to the president and to the NASA Administrator. (The White House Executive Order Establishing the Presidential Commission on the Space Shuttle Challenger Accident, Feb 3/86)

February 5: Further investigation into the Challenger accident opened a number of theories into the booster rocket's failure. One pointed to bad fuel segments, perhaps caused by cold temperatures. "Propellants can crack the way you crack a cookie. When they crack…it allows you new sites where gases can be generated," opined Herman Krier from the University of Illinois. Another theory suggested that improper handling had produced hairline cracks while fuel segments were cast or during launch preparation, or that installation of badly joined seams between the fuel segments caused uneven burning. It was noted that workers ignored standard operating procedures. A third theory looked into the possibility that a solid-rocket booster nozzle had burned through, as had nearly happened on a previous flight. Still another theory cited cold weather or vibrations resulting from launch as a possible culprit for a failure in the suspected seam that joined two fuel segments. Other theories speculated that the external tank experienced overheating, leading to expansion of its contents and rupture of the tank, or even direct burn-through from the 5,900 degrees of exhaust leaving the rocket booster or that there might have been a pre-existing hole in the tank. However the hydrogen came to leak, it was ignited by the heated air surrounding it or by the thin flame from the solid rocket.

NASA Engineers thought, in that point in the investigation, that cold weather had no adverse affect on either the solid fuel or the seals, but were concerned with the possibility of leakage between segments of the rocket motors because deterioration of these seals had been noted on previous flights. Back up seals, they added, performed adequately. (*P Inq*, Feb 5/86; Feb 10/86; *W Times*, Feb 5/86; Feb 6/86; Feb 7/86; *CSM*, Feb 5/86; *C Trib*, Feb 5/86; Feb 6/86; Feb 7/86; *NY Times*, Feb 5/86; Feb 7/86; Feb 9/86; *B Sun*, Feb 6/86; Feb 7/86; *WSJ*, Feb 6/86)

• The President's Space Shuttle Commission gathered in private before beginning its formal investigation into the disaster. Headed by former Secretary of State William P. Rogers, the Commission planned to weigh all of the evidence and make a recommendation, with assistance from NASA, after 120 days. (*USA Today*, Feb 5/86)

February 6: The President's Commission began probing into the Space Shuttle Challenger accident as it took statements from NASA officials Dr. William R. Graham, Acting Administrator; Jesse Moore, Associate Administrator; Arnold D. Aldrich, Space Shuttle Manager; Judson A. Lovingood, Deputy Manager; and Robert B. Sieck, Director, Shuttle Operations. As several possible reasons into the cause of the accident were explored, Moore emphasized that space flight would always be dangerous. (*NY Times*, Feb 7/86)

February 7: The White House accelerated its search for a permanent head of NASA to replace James M. Beggs, indicted in December on fraud charges unrelated to the agency. Beggs was on an unpaid leave of absence during the time of the Challenger accident; Deputy Administrator William R. Graham was Acting Administrator. Beggs was later acquitted of all wrong doing and the Justice Department apologized for the erroneous charges. (*W Post*, Feb 7/86; *Bus Wk*, Feb 17/86)

February 10: NASA reported that the United States and the United Kingdom had recently agreed to cooperate in the development of technology for an advanced short takeoff and vertical landing (ASTOVL) aircraft. The ASTOVL aircraft would have the advantages of an advanced supersonic fighter aircraft, but with the added capability of landing vertically. The cooperative research program would investigate four propulsion concepts: vectored thrust, ejector augmenter, tandem fan, and remote augmented lift system. The memorandum of understanding was signed by William R. Graham, NASA Acting Administrator, and Donald A. Hicks, Under Secretary for Research and Engineering, Department of Defense, for the United States, and by Sir David Perry, Chief of Defense Equipment Collaboration, for the United Kingdom. (NASA Release 86-10)

• NASA announced that it had postponed three launches: the Ulysses mission to investigate the poles of the sun; the Galileo mission to orbit Jupiter and send a probe into the planet; and the Astro-1 mission, an ultraviolet astronomy laboratory mounted in the Shuttle's payload bay that was to examine quasars, "hot" stars, galaxy centers, and Halley's Comet. The Ulysses and Galileo missions were scheduled for launch from a Shuttle in May, but both launches were dependent on certain relative positions of Jupiter and Earth, occurring only once every 13 months. The Astro-1 had to be launched in the March–April timeframe in order to observe Halley's Comet, and with that opportunity gone no new date was requested. The delays were caused by the January Space Shuttle accident. (NASA Release 86-11; *NY Times,* Feb 11/86)

• Advisors at NASA determined that future dependance on the Shuttle for all space operations would be contradictory to a reliable space transportation program. After several days of interviews, a 13-member panel, comprised of aviation and aerospace veterans, concluded that expendable launch vehicles should have greater consideration because the Shuttle was a research vehicle. (*W Post,* Feb 11/86)

• The Presidential Commission questioned NASA scientists regarding faulty o-rings as a cause of the Challenger accident. The panel learned that these seals could indeed malfunction in cold weather and also that erosion (not the result of cold weather) had occurred in 6 of 171 seals on previous flights. Morton Thiokol, the firm responsible for manufacturing the o-rings, had raised concerns about launching the Shuttle in cold weather, but NASA engineers concluded that the backup ring would function if the primary seal failed. Released NASA documents also showed concern expressed by engineers, who, three years prior to the accident, concluded that the mechanism for sealing segments of the solid rocket booster could fail and result in "loss of mission, vehicle, and crew." One risk analysis received by NASA from an outside agency estimated the chances for an accident involving the boosters as 1 in 35.

NASA's safety advisory board stated in their 1985 annual report that 18 Shuttle flights per year were "very optimistic" because these launches were anything but routine. However, "the requirement to launch has not been a prerequisite," said acting NASA Administrator Dr. William R. Graham, "The prerequisite has been safety." (Official Transcript Proceedings Before the Presidential Commission on Space Shuttle Challenger Accident, Feb 11/86; *P Inq,* Feb 12/86; Feb 13/86; *B Sun,* Feb 12/86; Feb 13/86; *CSM,* Feb 12/86; Feb 13/86; *C Trib,* Feb 12/86; Feb 14/86; Feb 13/86; *WSJ,* Feb 12/86; *USA Today,* Feb 12/86; Feb 13/86; *NY Times,* Feb 13/86; *W Post,* Feb 13/86)

February 12: Dr. William R. Graham, NASA's Acting Administrator, affirmed plans to continue NASA's Teacher in Space Program, administered by NASA's Educational Affairs Division under the direction of Robert Brown. The reason given: NASA's continuing goals were to "increase the prestige of the teaching profession, to increase the awareness in the education community of the impact of technology and science on this country's future, and to use space as a catalyst to enhance all subject areas and grade levels of our education systems." (NASA Release 86-12; *USA Today,* Feb 14/86; *NY Times,* Feb 14/86; *W Post,* Feb 14/86)

February 13: NASA released footage of the Challenger just seconds after ignition, showing a puff of black smoke coming from the right solid rocket booster near a joint between two segments. The tape lent more weight to the theory that the accident had resulted from a faulty seal in the motor. (*W Post,* Feb 14/86; *WSJ,* Feb 14/86; *W Times,* Feb 14/86; *NY Times,* Feb 14/86; *B Sun,* Feb 14/86; *USA Today,* Feb 14/86)

February 17: Forty miles off the Florida coast and at a depth of 1,200 feet, a four-man submarine recovered debris and took photographs of what engineers hoped was Challenger's right booster rocket. As the investigation moved further ahead, the Presidential Commission looking into the accident asked NASA to remove from its own investigation any officials involved in the decision to launch the Shuttle. (*B Sun,* Feb 18/86; *NY Times,* Feb 18/86; *W Post,* Feb 18/86; *P Inq,* Feb 18/86; *W Times,* Feb 17/86)

• NASA Requested $45 million for fiscal year 1987 to accelerate research for its next generation spacecraft, a hypersonic transatmospheric vehicle, or "space plane", capable of Mach 25 (about 17,000 mph). Able to fly into orbit, reenter the atmosphere and achieve orbit again while being completely reusable, the craft would have both commercial and military applications. (*W Times,* Feb 17/86)

February 19: Boeing Aerospace was selected to supply NASA with Shuttle upper stages to place two Tracking and Data Relay Satellites into geosynchronous Earth orbits. Engineering support and integration of the upper stages with the satellites were also to be provided. The contract was expected to run from March 1986 through April 1990 at a price of approximately $93 million. (NASA Release 86-14)

• The Presidential Commission's investigation of the Challenger accident, said the *Chicago Tribune,* had nearly concluded that cold weather led to reduced elasticity of the rubber seal in the rocket booster and the subsequent leak of gasses. Engineers from Morton Thiokol, the rocket booster manufacturer, objected to the cold weather launch, but officials higher up in that company gave NASA the go-ahead. (*NY Times,* Feb 20/86; *C Trib,* Feb 20/86; *W Post,* Feb 20/86)

February 20: Rear Admiral Richard H. Truly, commander of the Naval Space Command and former NASA astronaut, was appointed Associate Administrator for Space Flight. The new position called for Truly to head the Space Shuttle program and to direct NASA's Design and Data Analysis Task Force, which reviewed the Shuttle Challenger's January accident. He succeeded Jesse W. Moore in both of these roles. (NASA Release 86-15; *WSJ*, Feb 20/86; *NY Times*, Feb 20/86; *W Times*, Feb 21/86; *B Sun*, Feb 21/86)

• The Soviet Union launched what it said was a "third generation" Space Station named Mir, the Russian word for peace. The Mir was larger than their Salyut-7 Space Station and had six docking ports for habitable modules. With the addition of Salyut laboratories, Russia would have an orbiting complex large enough for permanent crew habitation. (*B Sun*, Feb 21/86; *W Times*, Feb 21/86; *W Times*, Mar 4/86; *CSM*, Mar 7/86)

February 22: Thomas L. Moser, Director for Engineering at NASA's Johnson Space Center, was appointed as Deputy Associate Administrator for Space Flight to investigate the Space Shuttle Challenger accident. (NASA Release 86-16)

February 24: A NASA study found that the part of the booster rocket that was suspect in the Challenger accident had reached a temperature of only about 29 degrees Fahrenheit at the time of liftoff, 9 degrees cooler than the surrounding air. Both NASA and Morton Thiokol engineers, the study noted, had knowledge of the ring's susceptibility to damage in 50-degree temperatures. A member of the President's Commission pointed out that NASA officials should have been aware of this as earlier temperatures were in the 20s and the o-rings would not heat up quickly. (*W Post*, Feb 24/86)

February 25: James M. Beggs resigned as NASA's Administrator. Candidates President Ronald Reagan was considering for the job included James C. Fletcher, Thomas O. Paine, James A. Abrahamson, Frank Borman, Harrison Schmitt, and Lew Allen. (*NY Times*, Feb 25/86; *WSJ*, Feb 26/86)

• Engineers from Morton Thiokol gave testimony before the President's Commission that the decision to launch the ill-fated Shuttle was a management initiative, and they suggest that those people had ignored o-ring safety data. The engineers testified that their decision-making process was reversed as the burden of proof shifted and was now on them to show that the launch would be unsafe. Contrary to what a number of engineers stated to the Commission, one management official testified that there was no undue pressure from NASA to go ahead with the launch and that management's decision was based solely on available data. (*B Sun*, Feb 26/86; *NY Times*, Feb 26/86; *W Post*, Feb 26/86; *W Times*, Feb 26/86)

• NASA's acting administrator Dr. William R. Graham announced that he would not hesitate to make substantial changes wherever they might be warranted in the organization. Graham said, "That goes not only for the [launch] decision-making process, but also for engineering, design, vehicle certification and qualification." (*W Times*, Feb 26/86)

February 26: As further testimony was heard by the President's Commission, NASA manager Larry Mulloy stated that objections made by Morton Thiokol engineers to halt liftoff seemed illogical, but added that had he known that all of the company's engineers held the opinion, he might have changed his mind. Other testimony given by NASA officials conceded that they argued against the idea of cold weather being detrimental to a launch, but had little data to demonstrate this, and insisted that this opposition in no way pressured Morton Thiokol management to reverse the recommendation of its engineers. (Official Transcript Proceedings Before the Presidential Commission On Space Shuttle Challenger Accident, Feb 26/86; *NY Times*, Feb 27/86; *C Trib*, Feb 27/86; *LA Times*, Feb 27/86; *W Post*, Feb 27/86; *P Inq*, Feb 27/86; *W Times*, Feb 27/86; *WSJ*, Feb 27/86)

• The Challenger accident forced NASA to reconsider expendable launch vehicles as a means for putting satellites into space. Believed to have been rendered obsolete by the Shuttle, these rockets were once again deemed necessary for meeting the growing schedule of launches. The change of position notwithstanding, NASA and the Department of Defense appeared ready to request another orbiter. (*B Sun*, Feb 27/68)

February 27: The United States Space Foundation outlined plans to provide a focal point that would raise funds for a replacement of the Space Shuttle Challenger. The Challenger 7 Fund reached out to individuals and organizations for contributions while hoping that Congress would authorize the replacement. (NASA Release 86-18)

• Top officials at Rockwell International, builder of NASA's orbiters and closely tied to the Shuttle program, testified that they believed cold weather had made launch of the Challenger unsafe. The President's panel concluded that NASA's launch decision process was flawed because criteria for approval should not have been changed at the last minute. The *New York Times* concluded, "The picture emerging from the inquiry into the loss of the Space Shuttle is one of chronic failure in the space agency and its contractors to communicate life-and-death problems up the chain of command." (*W Post*, Feb 28/86; *NY Times*, Feb 28/86; *C Trib*, Feb 28/86; *WSJ*, Feb 28/86)

February 28: The Inter-Agency Consultative Group (IACG) on Halley's Comet encounter was scheduled to meet March 4–9 in Moscow and March 12–14 in

Darmstadt, Federal Republic of Germany. Formed in 1981, the IACG comprised representatives from NASA, the European Space Agency (ESA), the Soviet Union's Intercosmos Council, Japan's Institute for Space and Astronautical Studies (ISAS), and the International Halley Watch.

ESA's *Giotto,* the Intercosmos Council's VEGA 1 and 2, and the ISAS Suisei (Comet) and Sakigake (Pioneer) spacecraft were launched in the previous 15 months to observe the Comet's passage around the Sun. NASA's Launching of a Spartan UV telescope and an Astro UV telescope to observe the Comet from Earth orbit was terminated because of the Challenger accident. But NASA's Pioneer-Venus spacecraft (launched in May 1978) had already completed five of its seven planned weeks of viewing the Comet, and the International Cometary Explorer spacecraft (launched in August 1978) was scheduled to monitor Halley's solar wind upstream. (NASA Release 86-19; *P Inq,* Feb 28/86; *NY Times,* Mar 4/86)

March

March 1: James R. Thompson, of Princeton University's Plasma Physics Lab, nationally recognized propulsion expert, and former NASA propulsion engineer, was named vice chairman of the NASA task force inquiring into the Challenger accident. His appointment was in support of the Commission assigned by President Ronald Reagan to perform the investigation of the Challenger accident. (NASA Release 86-20)

March 3: NASA announced a comprehensive project to evaluate the implications of the Challenger accident on the space program. Because the Ulysses and Galileo missions had been postponed, NASA announced that the Shuttle Discovery would be modified so that it could launch spacecraft with a Centaur upper stage rocket. The Challenger accident left only the Shuttle Atlantis with the capability for carrying Centaur payloads.

These decisions were made by the Acting NASA Administrator, Dr. William R. Graham, on the recommendation of the Headquarters Replanning Task Force, headed by Dr. Raymond S. Colladay, set up to study program alternatives in the wake of the accident. Other matters considered by the task force included: (1) the requirements for all aspects for an orbiter to replace the Challenger as well as the lost Inertial Upper Stage (IUS) launch vehicle and its support structure attached to a Tracking and Data Relay Satellite (TDRS) lost with the Shuttle; (2) new launch schedules for the remaining three Shuttles; (3) an additional launch site at Vandenberg Air Force Base; (4) the role of expendable launch vehicles for commercial use; (5) the retention of ground communications because only one TDRS, instead of two, was in orbit to link spacecraft to earth because of the accident; (6) and identification of monetary losses other than those from the destroyed equipment.

In the meantime, customers who had hoped to launch their satellites from the Space Shuttle were being forced to examine other alternatives. With launch schedules certain to fall far behind and military and scientific communities getting first priority, commercial interests would have to either put their projects on the back burner or use NASA's rival, the European Space Agency's Arianespace company, for more expensive launches. With NASA's announcement that its Shuttle program would be put on hold for 12–18 months, Shuttle–linked businesses were searching for other options as well. (NASA Release 86-22; *C Trib,* Mar 3/86; *B Sun,* Mar 4/86; *USA Today,* Mar 4/86)

• A lightweight transmitter for sending messages to weather satellites was placed on polar bears and caribou in the spring of 1985, and had so far functioned very well, said biologist Steven Amstrup. Because of the satellite's ability to track these animals, biologists could determine whether the animals were walking, running, sleeping, or foraging for food. (*B Sun*, Mar 3/86)

• A device conceived by Donald Young of NASA's Ames Research Center, Mountain View, California, for aiding scientists in treating bone loss resulting from extended space flights, spawned hope among the medical community. The vibrating analyzer is placed in the center of a limb and the ratio of force applied to the bone and the bone's displacement indicate the stiffness of the bone. The device, it was thought, would offer earlier detection of osteoporosis and take guess work out of cast removal. (*WSJ*, Mar 3/86)

March 4: NASA's chief engineer said that cutbacks in the safety program over the previous three years had put the Agency "in a dangerous situation." Loss of personnel and transfer of safety responsibility to field offices and contractors contributed to NASA's inability to catch the seal problem. He reasoned that "there are only so many problems [one person] can look at, and in my opinion the solid rocket booster was a relatively easy thing to do. But it really wasn't getting the amount of attention from here it should have." (*W Times*, Mar 5/86)

March 5: NASA and the Nuclear Safety Review Panel differed in their estimates of the reliability of booster rockets before the January 28 accident. The interagency panel's estimate of failure was 1 in 1,000 launches as compared to NASA's 1 in 100,000 firings. The panel was created because 19 missions carrying potentially hazardous plutonium power sources (were an accident to occur within the Earth's atmosphere) involved the Shuttle. Two other nuclear powered satellites were scheduled for launches from the Shuttle later in 1986. (*WSJ*, Mar 5/86; *C Trib*, Mar 6/86)

• Aerojet Solid Propulsion Company, an unsuccessful bidder on Shuttle boosters, released documents it had sent to NASA some 13 years earlier stating that the multi-segment booster that relied on o-ring seals were "burdened with design features that can detract from safe, efficient, and reliable operation." According to officials involved in the decision to choose the Utah based Thiokol design, the least expensive bid; the fact that Senator Frank Moss, then chairman of the committee that controlled NASA's budget came from Utah; and that then NASA Administrator Dr. James Fletcher was former president of the University of Utah were all salient factors. NASA calculations, however, also pointed out that Thiokol needed minimal up front Government cash and deferred major costs to the latest possible date, and that Aerojet required a good deal more money to get started. All three factors were important to the early days of the program's budget constraints. (*B Sun*, Mar 5/86)

March 6: Several of the more than 30 U.S. scientists serving as coinvestigators on the European Space Agency's Giotto mission were members of the television team scheduled to analyze and televise images of Halley's Comet from the spacecraft on March 13. Launched in July 1985, the Giotto was one of five spacecraft (others were the Soviet Union's Vega 1 and 2 and Japan's Suisei and Sakigake) headed for an encounter with the comet. In position 300 miles from Comet Halley's nucleus, the Giotto promised to obtain the highest resolution imaging. (NASA Release 86-23)

• Rear Admiral Richard H. Truly, Associate Administrator for Space Flight, announced assignments to the NASA 51-L Data and Design Analysis Task Force. The task force was to collect and analyze information to support a thorough review of all aspects and potential causes of the Challenger accident. (NASA Release 86-24)

• NASA announced that U.S. scientists would have the opportunity to participate with the Institute of Space and Astronautical Science (ISAS) of Japan in the High Energy Solar Physics program. The goal of the mission was to better understand high energy phenomena on the Sun through x-ray and gamma ray observation instruments carried on a spacecraft. (NASA Release 86-25)

• NASA officials provided investigators with new evidence challenging the theory that the Challenger accident was triggered by failed rocket seals caused by cold weather. They presented pre-launch photographs showing a flaw in the suspected o-ring on the right side booster, a possible correlation to the fact that workers assembling the booster were hampered with a misshaped rocket segment. New tests on the o-rings conducted at Morton Thiokol's plant demonstrated that they sealed properly to temperatures at 10 degrees below zero. Also, the Challenger, according to meteorologist Irving P. Krick, sustained wind shear of 125 to 150 miles per hour just prior to the accident, which would have been similar to "flying into a tornado." (*W Post,* Mar 7/86)

• The Soviet Union's Vega I spacecraft sent back pictures of Halley's Comet as it passed within 5,500 miles of its nucleus. The photographs, displayed on screens at a Soviet Space Center, were viewed by scientists from around the world and brought the scientific community together behind the Iron Curtain. A host of new data such as the possibility that the Comet contains two nuclei, that the core is three or four miles across, and evidence bolstering the idea that the Comet is leftover debris from the solar system's creation some 4.6 billion years ago was revealed. The presentation also gave American scientists a chance to study Halley's Comet; the United States launched no spacecraft for that purpose. The Soviet's second probe, Vega II, passed 5,125 miles from the comet's core on March 9. (*FBIS,* Tass (Eng trans), Mar 7/86; *NY Times,* Mar 7/86; *W Post,* Mar 7/86; *W Times,* Mar 7/86; *B Sun,* Mar 7/86; *C Trib,* Mar 7/86; Mar 10/76; *CSM,* Mar 7/86; March 10/86)

March 8: The Japanese spacecraft, Suisei, passed within 94,000 miles of Comet Halley's nucleus on this date. Although the pass was far more distant than passes by the Soviet's two Vega spacecraft, or the European Space Agency's Giotto, Japanese scientists were immensely pleased with the data. (*NY Times,* Mar 17/86)

March 9: NASA announced that it had located both the Shuttle's crew compartment and the remains of some crew members, in 100 feet of water, about 25 miles northeast of Cape Canaveral, Florida. Sonar first detected the object on the night of March 7 and subsequent dives confirmed it to be the module. About 10 percent of the wreckage had been recovered to this date, but the suspected right solid rocket booster still lay on the ocean's floor. Although little insight into the cause of the accident was expected, recovery of the data tapes, officials speculated, could shed some light. (*USA Today,* Mar 10/86; *P Inq,* Mar 10/86; *B Sun,* Mar 10/86; *NY Times,* Mar 10/86; *W Post,* Mar 10/86; *W Times,* Mar 10/86)

March 11: Dr. William R. Graham told a congressional committee that the suspected o-rings in the Challenger accident should be modified or redesigned. The manufacturer of the rocket boosters had submitted 43 possible modifications to NASA since February, and changes to the fleet, said Graham, would cost about $350 million. This was the first time any official from NASA had indicated a need for improving the seals since the January 28 accident.

The acting administrator also noted that recovery from the accident could cost the Nation $3.2 billion and replacing the Challenger would run another $2.8 billion. Not replacing the Shuttle, he warned, would restrict the fleet to military use and service flights. (Hearings Before the Subcommittee on Space Science Applications, Mar 11/90; *P Inq,* Mar 12/86; *W Post,* Mar 12/86; *USA Today,* Mar 12/86; *W Times,* Mar 12/86)

March 12: Bernard M. Oliver, Chief of the Search for Extraterrestrial Intelligence (SETI) program at NASA's Ames Research Center in Mountain View, California, received the National Medal of Science from President Ronald Reagan. The award recognized his accomplishments while working for Bell Telephone Laboratories, where he helped establish standards for television transmission, worked on automatic radar tracking, and advocated pulse-code modulation for the telephone system; and his accomplishments at Hewlett-Packard, where he developed the first hand-held calculator with scientific functions and SETI, a program designed to search the stars for radio signals. (NASA Release 86-27)

• The Congressional Budget Office estimated that it would cost $5 billion to both replace the Challenger Shuttle and build more expendable launch vehicles for carrying payloads into space. NASA had planned to make all of its satellite launches from an orbiter, but changed its position after the Challenger accident. The shift was welcome news to the rocket industry, even though the companies would need two years to get ready for launch inasmuch as the rocket program, with no crew assistance, had been all but replaced in the United States by the Shuttle program. (*NY Times*, Mar 13/86; *W Post*, Mar 13/86)

• NASA and Morton Thiokol found new evidence suggesting that the putty used in joints of the solid rockets to protect the o-rings could have stiffened from the cold and prevented the o-ring from sealing. They also noted that the putty could have cracked and exposed the o-ring to hot gases. Adequate data measuring the effect of cold on the putty did not exist. Dr. William Graham notified the Presidential Commission of these findings. (*WSJ*, Mar 13/86)

• Salvage crews reported that they might have located a key part of the booster rocket, suspected in the January 28 accident, 32 miles offshore at a depth of 650 feet. Embedded firmly into the ocean's floor, the estimated 400–500 pound debris located near other wreckage was positively identified as parts from the right booster rocket. As engineers examined their most recent discovery, other teams recovered still more of the crew's cabin as well as additional remains and personal affects. (*W Times*, Mar 13/86; *NY Times*, Mar 13/86; *B Sun*, Mar 13/86)

• The former Soviet Union launched two veteran cosmonauts into orbit, scheduled to arrive at the Mir Space Station after two days. The launch, contrary to former Soviet practice, was both announced and televised. The launch, said James Oberg, author of several books about the Soviet space program, was done "with great propaganda benefit." In an effort to demonstrate their achievements, he stated further, "They have a greater self-confidence in their program [and] it helps concentrate world attention on the benign part of their space program." U.S. experts acknowledged that Soviet Space Station technology was years ahead of us and that the mission might well begin man's permanent occupation of space. (*FBIS*, Tass (Eng trans), Mar 12/86; *B Sun*, Mar 14/86; *P Inq*, Mar 14/86; *W Times*, Mar 14/86; *NY Times*, Mar 14/86; *W Post*, Mar 14/86; *C Trib*, Mar 14/86)

• The European Space Agency's spacecraft Giotto encountered Halley's Comet as it passed within 335 miles of the body's nucleus. The craft transmitted pictures of the comet every four seconds up until two seconds and 930 miles away, when its camera was damaged by dust particles traveling 50 times the speed of a bullet. These were the closest pictures by far taken of the Comet. Multinational scientists at the West German control center were jubilant about data gathered by sophisticated instruments. The mission was the most spectacular taken on by the European Space Agency, and the international cooperation among scientists was lauded by the media. (*B Sun*, Mar 14/86; *NY Times*, Mar 14/86; *W Post*, Mar 14/86; *W Times*, Mar 14/86)

March 14: Dr. Albert Boggess was awarded the Herschel Medal, by the Council of the Royal Astronomical Society of the United Kingdom, for his contribution to the success of the International Ultraviolet Explorer (IUE). The IUE satellite was designated to study Comet Halley's evolution as it approached and receded from the Sun. Boggess was named co-recipient of the award with University of London Professor R. Wilson. (NASA Release 86-28)

March 16: This date marked the 60th anniversary of liquid-fueled rocket flight. In 1926, Dr. Robert H. Goddard assembled and launched a gasoline propelled rocket from his aunt's farm in Worcester, Massachusetts. In commemoration of the event, NASA planned to launch a full-scale replica of the rocket from the Goddard Space Flight Center, Greenbelt, Maryland. (NASA Release 86-26)

• A 3,500-pound piece of debris that was determined to be a rocket booster, but not identified as from the left or right side, was recovered 28 miles northeast of Cape Canaveral, Florida, in 400 feet of water. The massive operation was resumed after a two-day halt from storms and rough seas. Other salvage teams also continued their efforts to recover the rest of the crew's compartment and remains. (*W Post,* Mar 17/86; *P Inq,* Mar 17/86; *NY Times,* Mar 18/86)

March 18: NASA's orbiting telescope, International Ultraviolet Explorer, observed Halley's Comet as it encountered European (Giotto), Soviet (Vega 2), and Japanese (Suisei) spacecraft. The September 1985 to March 1986 observations of the Comet were made with hopes of granting scientists a better understanding into its long-term behavior. The telescope measured the Comet's water ejection rate, and its carbon, sulfur, and oxygen content, as well as its variation of visual brightness. The advantage of observations from a telescope as opposed to those from a spacecraft, said IUE telescope operations manager George Sonneborn, was the much wider angle of view it offered. (NASA Release 86-29)

• NASA and the Science and Technology Agency of Japan announced an agreement reached on the hardware that Japan will carry into Phase B Space Station definition and preliminary design. According to the agreement, the preliminary design activities were to include: (1) a pressurized module to provide shirtsleeve work space for station crews; (2) an exposed work deck; (3) a scientific/equipment airlock; (4) a local remote manipulator arm; (5) and an experiment logistics model. The multipurpose laboratory called for accommodation of general scientific and technology development research, including microgravity research, as well as control panels for operating the Space Station's mobil remote manipulator system and attached payloads. (NASA Release 86-31)

• NASA released a list containing 748 critical parts of the Space Shuttle that could lead to disaster if they did not function properly. All of these components lacked a backup system and were therefore placed in the "Criticality 1" category. Unlike the fuselage or heat-shielding tiles, for example, 617 of the critical parts could be covered by backup equipment, but engineers waived the measure as they felt confident that failure of them was unlikely. Arnold D. Aldrich, Manager, National Space Transportation System, said, "We are reviewing in detail all the items on the Critical Items list." (Nat Sp Trans Sys: Critical Items List; *W Post,* Mar 18/86; *W Times,* Mar 18/86; *B Sun,* Mar 18/86)

March 18: James R. Thompson, vice-chairman of the NASA task force investigating the Challenger accident, announced that the cause of the disaster would be pinpointed within a month. Although the right booster rocket, which would be a tremendous help, was still unrecovered, and tapes from the crew compartment shed no light, Thompson noted that photographic evidence and other data already available would eventually yield the answer. He also stated that April 18 was the deadline given by President Ronald Reagan's panel for overall investigation, with whom, he added, NASA investigators were working harmoniously. (*USA Today,* Mar 19/86; *W Post,* Mar 19/86; *NY Times,* Mar 19/86; *P Inq,* Mar 19/86; *C Trib,* Mar 20/86)

March 19: East Coast winter storms in the form of cyclones were studied by scientists from NASA's Goddard Space Flight Center, Greenbelt, Maryland. Concentrating their efforts on costal regions of North and South Carolina, where these storms develop and move northward, NASA scientists analyzed the upper atmosphere conditions using moisture and wind sounding balloons and the Nimbus 7 satellite to collect data.

Another criterion for understanding these storms was the analysis of the transfer of moisture from the ocean into the atmosphere. NASA flew its four-engine Electra aircraft, outfitted with a combined laser/telescope (Lidar), to detect salt spray and aerosols as it fired a lazar beam that allowed the telescope to measure rising moisture and convection patterns. NASA's ER-2 aircraft was used in higher altitudes for the same purpose; it carried microwave radio-meters able to detect water vapor and cloud liquid water content. (NASA Release 86-30)

• NASA scheduled five expendable launches in 1986 from Cape Canaveral, Florida, employing Delta rockets, used in joint management with the McDonnell Douglas launch team, and Atlas Centaur rockets, jointly managed launches with General Dynamics Convair. Of the five satellites, two were for the National Oceanic and Atmospheric Administration. The Geosynchronous Operational Environmental Satellites were designed for weather forecasting and collecting data on various environmental effects. The other three were for the Department of Defense. Two Fleet Satellite Communication satellites would provide secure communications between land-based facilities and ships, submarines, and aircraft, and one other was for support of the Strategic Defense Initiative. (NASA Release 86-32)

March 20: The U.S. Government considered plans for replacing the Challenger by allowing private investors to raise the money and lease the new Shuttle to NASA, stretching out payments for perhaps 15 years. The plan called for NASA to order a replacement shuttle from Rockwell, but Willard Rockwell's new firm, Astrotech, would buy it. Furthermore, investors would be reimbursed by NASA for any accident loss, and Astrotech would market the launches to commercial customers. (*WSJ,* Mar 20/86; *W Post;* Mar 30/86; *B Sun,* Apr 1/86)

• Debris brought to shore on March 19 was determined to be part of the right booster rocket from the Challenger accident. The white outer skin of the pieces were discolored from excessive heat, and the suspected o-ring, officials said, may lie inside of one 4-foot-by-5-foot piece of wreckage. (*W Times,* Mar 21/86; *USA Today,* Mar 21/86)

March 21: Canadian Prime Minister Brian Mulroney announced Canada's decision to proceed with Space Station participation when he visited with President Ronald Reagan. Canada agreed to perform preliminary design of a Mobile Servicing Center during Phase B of the Space Station definition and preliminary design study that would consist of a base structure with accommodations for payloads, orbital replacement units, utilities and thermal control. It would function as a multipurpose structure equipped with manipulator arms that would be used to help assemble and maintain the Space Station, as well as help keep instruments and experiments mounted on the Station's framework. (NASA Release 86-33)

• NASA officials concluded, and reported to the President's Commission, that the o-rings in Challenger's right solid-rocket motor were indeed the cause of the January 28 accident. They were still unable to cite a reason for failure of the seals, but hoped that tests over the next 10 days would give some clue. NASA also noted that the putty used to protect these seals might shed some light. (*WSJ,* Mar 24/86; *CSM,* Mar 24/86)

March 24: NASA planned to shift its original plan of carrying a mixture of scientific, military, and commercial payloads and give greater emphasis to the military when Shuttle flights resumed. The military, they added, always had top priority and five to seven of the flights slated for next year were devoted to them. (*NY Times,* Mar 24/86)

• A Presidential panel report was to be released April 11, but details that aimed at mining the Moon and asteroids and establishing human presence on Mars were disclosed in *Aviation Week & Space Technology,* a trade magazine. The plan, which assumed that the Space Station would be completed by 1994, called for establishing a lunar base after the year 2000 for mining and production of rocket fuel, expanded searches for potentially useful asteroids, and a "network of spaceports between Earth, the Moon, and Mars and a Martian colony by the year 2027." (*W Post,* Mar 24/86; *NY Times,* Mar 25/86; *CSM,* May 23/86; *P Inq,* May 23/86)

March 25: NASA's Jet Propulsion Laboratory in Pasadena, California, selected the RCA Corporation and Orbital Sciences for negotiations leading to the award of contracts to build a spacecraft and upper stage booster for the Mars Observer Mission scheduled for launch in August 1990. The first in a series of proposed planetary observer programs, the Mars Observer would study the climate, atmosphere, and surface, using eight science instruments while in orbit around the planet a full Martian year, 687 Earth days. (NASA Release 86-34; *WSJ,* Mar 26/86)

• Speaking to NASA employees, Shuttle Director Richard H. Truly announced that the Shuttle Program would resume in about one year, after an intense safety program to include reassessment of Shuttle management structure, redesign of the solid rocket boosters, reassessment of methods for aborting Shuttle flights, and recertification of all critical Shuttle parts. He noted that "the business of flying in space is a bold business," and added, "We cannot print enough money to make it totally risk-free. But we certainly are going to correct any mistakes we have made. . .and we're going to get going ahead just as soon as we can."

When the Shuttle Program resumes, he continued, the first Shuttle would be launched from Florida, not have a guest astronaut, the payload would be a familiar one, and the Shuttle would land at the better landing strip at Edwards Air Force Base, in California. (*USA Today,* Mar 26/68; *W Post,* Mar 26/86; *WSJ,* Mar 26/86; *NY Times,* Mar 26/86; *W Times,* Mar 26/86)

March 26: An official from the Pentagon warned members of Congress that a one-year delay in the Shuttle launches would create a backlog of 10 military missions and a 2-year delay would raise that number to 21. He and others viewed this as a "national emergency." The Air Force Undersecretary further pointed out a need for a replacement Shuttle and 10 additional expendable launch vehicles. Congressional committees, however, had not yet decided how the cost, if approved, would be shared by the Air Force and NASA. (*B Sun,* Mar 27/86; *W Times,* Mar 27/86)

March 28: James E. Kingsbury, director of the Science and Engineering Directorate at NASA's Marshall Space Flight Center, in Huntsville, Alabama, was named to manage an *ad hoc* group formed to requalify the motor of the Space Shuttle's solid rocket booster for flight. The booster rockets, said NASA, could be ready in one year. Kingsbury said that all of the current hardware could be modified and re-used, a comment that surprised outside engineers. (NASA Release 86-36; *W Post,* Mar 28/86; *NY Times,* Mar 28/86; *NY Times,* Mar 30/86)

April

April 1: A student photography experiment, "CAN DO," from the Charleston County School District, South Carolina, scheduled to fly on the Space Shuttle Columbia March 6 was rescheduled for April 5-24. The goal of the experiment was to observe Comet Halley from NASA's high-flying Kuiper Airborne Observatory aircraft, a converted C-141. The aircraft was scheduled for several high-altitude flights over New Zealand, a part of the southern hemisphere where the Comet would be most visible.

Students from other schools, both in the United States and abroad, were to participate in the program by photographing the Comet from the ground and comparing their pictures with those taken in flight. The pictures taken at ground level would assist professional astronomers because most large observatory telescopes are not designed to view large objects, and resolving the fine detail in Comet Halley's tail could be enhanced with the much wider overall view from a standard camera lens. (NASA Release 86-35)

• A section of debris recovered from the Atlantic Ocean in March was found to be the part of the right booster rocket suspected as the cause of the Challenger accident. Although the piece was a section opposite the flame seen coming from a joint in the rocket, it contained part of the joint. Recovery of the section could both aid investigators and assist engineers in their redesign. (*B Sun,* Apr 2/86; *W Post,* Apr 2/86; *NY Times,* Apr 2/86)

April 2: In a data gathering activity in support of the Challenger accident investigation, destacking of solid rocket motors to gather information and assess the pre-flight conditions in and around the field joints was scheduled for early April. Inspections would focus on the field joint o-rings, putty used in the joints, rocket motor case ovality or "roundness," with any assembly damage, clevis gap-opening resulting from stacking, and the degree of propellant slumping that occurred as a result of vertical stacking. Any relevant data from the destacking inspections that might be useful to the investigation was to be assembled by the NASA 51-L Data and Design Analysis Task Force and provided to the Presidential Commission on the Space Shuttle Challenger Accident. (NASA Release 86-37)

• NASA and the Astronauts Memorial Foundation, Inc. announced plans to build a memorial at the Kennedy Space Center in Florida, dedicated to astronauts who lost their lives while flying, training, or awaiting assignment to fly for the space agency. The memorial would honor the 1986 Space Shuttle

Challenger crew—Francis Scobee, Michael Smith, Judith Resnik, Ronald McNair, Ellison Onizuka, Christa McAuliffe, and Gregory Jarvis; the 1967 Apollo crew—Virgil Grissom, Edward H. White II, and Roger Chaffee; Charles Bassett and Elliott See, killed in 1966, Theodore Freeman killed in 1964, and Clifton Williams killed in 1967—all killed in T-38 trainer aircraft accidents; and Edward Givens, killed in 1967 in an automobile accident. (NASA Release 86-38)

April 3: The Presidential Commission investigating the Challenger accident recommended major changes in the Shuttle Program. Recommended changes included an injection system for astronauts, even though the Commission concluded that such a system would not have saved lives in the January 28 accident; and an independent safety board with power to postpone any flight in which potential hazards were recognized. Other problems pointed out by astronauts and NASA officials who testified were a faulty brake design, a rough and narrow runway at Kennedy Space Center that became more treacherous with Florida's unpredictable weather, a lack of spare parts, and deficiencies in spacecraft simulators.

NASA manager Arnold Aldrich also testified before the Commission and said that major flaws existed in the Shuttle Program. He was referring to communication problems that kept him uninformed about numerous matters, the most important of which were issues raised about the safety of seals in the rocket boosters. (*W Post,* Apr 4/85; *NY Times,* Apr 4/86; Apr 7/86; *USA Today,* Apr 4/86; *P Inq,* Apr 4/86; *WSJ,* Apr 4/86; *W Times,* Apr 4/86; *C Trib,* Apr 4/86)

• The *Wall Street Journal* reported that a Reagan administration interagency group would soon recommend that NASA discontinue commercial and foreign launches. NASA objected to the idea because of the likely permanent loss of the business to foreign competitors. The group also recommended a replacement for the Shuttle Challenger and a greater supply of expendable launch vehicles, but did not suggest how these costs could be met. (*WSJ,* Apr 3/86)

April 4: Space Shuttle Columbia was scheduled to be ferried atop a modified Boeing 747 and transported to Vandenberg Air Force Base in California where it was expected to remain until early November. Testing would closely parallel that done at Kennedy Space Center (KSC), in Florida, in preparation for the first West Coast Space Shuttle launch. Upon arrival, Columbia would be moved to the launch pad and mated with a set of solid rocket boosters and an external tank for integrated vehicle testing, including the loading of cryogenic propellants in a "wet" countdown demonstration test. Prior to the move, processing at KSC would include installation of the main engine, auxiliary power units, and orbital maneuvering system pods, in parallel with structural inspections, approved modifications, and ferry flight preparation. (NASA Release 86-40)

April 7: NASA and the Pentagon announced the award of $500 million in contracts to seven different defense contractors for research into the development of the aerospace plane, seen by many as a successor to the Shuttle. The craft was envisioned as one that could take off and land like an airplane, but have the added capability of achieving orbit where it could then travel at 25 times the speed of sound. (*WSJ*, Apr 8/86; *USA Today*, Apr 8/86; *W Post*, Apr 8/86)

• The Soviet Union invited foreign correspondents into its mission control center for a conference concerning orbiting cosmonauts and a live television interview with them. The Soviet Union set still another precedent when it allowed coverage of its space program by a large group of foreign press. The invitation followed a pattern of a more open Russian space agency, the televising of the recent liftoff of the cosmonauts on March 13, and the opening of the Soviet Space Research Institute to foreign visitors when the Vega spacecraft approached Halley's Comet were also first time events. (*W Post*, Apr 8/86; *W Times*, Apr 8/86)

April 8: James R. Thompson, vice chairman of the NASA task force investigating the Challenger accident, said that NASA officials could be "walking on the edge of a cliff" if they did not correct flaws in the solid rocket booster. Previous problems, he noted, simply were not taken seriously enough; he then added that NASA could say conclusively that the right solid rocket booster had caused the Shuttle accident. The task force, said Thompson, focused on four factors of the joint in the booster rocket: the tendency of the joints to rotate during liftoff; the impact of sub-freezing temperatures on the o-rings; defects in the putty used to protect them; and slight damage to the o-rings during assembly. James Kingsbury, director of Science and Engineering Directorate disagreed with Thompson's findings because the rockets had performed well on 24 previous flights. (*USA Today*, Apr 9/86; *W Post*, Apr 9/86; *P Inq*, Apr 9/86)

April 9: NASA's Task Force on the Scientific Uses of the Space Station, established in March 1984 and chaired by Stanford University Professor Peter M. Banks, released its second major report for planning the scientific utility of the Space Station. The report pointed out that space-based scientific progress had been slower than planned because of short time in orbit, rigid time lines, long periods between flights, and that the Space Station had the obvious potential of overcoming these difficulties.

The two major conclusions reached noted that Space Station facilities must be operated with the goal of producing outstanding scientific results; and that the station would be judged with respect to cost and research capability and the need for well-equipped, permanently habitable laboratories able to support a broad range of fundamental research in space.

The task force recommended that the Space Station emulate the adaptive science methodology used in terrestrial laboratories, introduce "telescience" (the ability to conduct research remotely), review safety standards to achieve a reasonable state of personal and system security, work with a crew size of 10, and develop "space mail" for delivering small samples of materials to ground laboratories. Other recommendations: (1) Attached payloads are an important part of the core Space Station, and NASA needs a more productive plan for converting observation experiments developed for Spacelab to Space Station; (2) free-flying platforms are essential for conducting many important scientific endeavors for the Space Station era (i.e., several platforms operating in different orbits); (3) NASA should enhance biological research activities in the pre-initial orbit configuration period using Spacelabs and other attached payloads on the Space Shuttle to gain experience for conducting biological science research programs; (4) development of a human-tended mode of scientific activity aboard a Space Station would be of little value and a need exists for a truly long-term, human capability; (5) it is essential that NASA look ahead to the activities that are anticipated over the 25–30 year life span of the core facility and its associated elements; (6) the Space Station will facilitate development of a new type of research termed "science in space"; (7) the science operations of the Space Station should be separated from the operational management of the overall facilities; and (8) NASA must be prepared to change many of its nationally oriented selection, funding, and management procedures because of the important hardware and scientific contributions that will be made by international partners. (Written Statement to the Subcommittee on Space Science and Applications, Committee on Science and Technology from Peter M. Banks, Apr 9/90; NASA Release 86-43)

• Evidence revealed the Challenger crew cabin did not explode with the mid-air accident, but remained intact until it struck the water. The cabin, said engineers, was seven times stronger than other parts of the Shuttle because it had to maintain an Earth-like atmosphere while in space. All seven astronauts, however, were probably killed instantly by shock from the initial blast, sudden depressurization of the cabin, or the tumbling nine-mile descent. (*W Post,* Apr 10/86; *NY Times,* Apr 10/86; *W Times,* Apr 10/86; *CSM,* Apr 10/86; *USA Today,* Apr 10/86; *C Trib,* Apr 10/86)

April 14: The United States planned to resume its efforts to put nuclear reactors into space. The move would renew a program abandoned 10 years prior to this date, but now deemed necessary for President Ronald Reagan's Strategic Defense Initiative (SDI), "Star Wars." Even without SDI, said Air Force officials, reactors would be needed for the next generation of space-based weapons; NASA could use such reactors to power Space Stations, lunar bases, interplanetary explorers, and satellite traffic control. The United States has mainly relied on solar energy for its space activities, but SDI would require more power than the panels could give. The Soviet Union never abandoned its plan for space reactors and sent up 20 nuclear-powered satellites, one of which crashed and spread radioactive debris over northwest Canada. (*W Post*, Apr 14/86)

• NASA officials announced that the redesign of booster rocket seals would render them fail-safe. The new design would not use putty to protect the seals from heat; would include an outside cover to protect the seals from rain and ice, with a possible heating element built in; use a clip to prevent rotation; and provide a backup seal for each joint of the booster. (*P Inq*, Apr 14/86)

• Rocket debris recovered on April 13, 40 miles northeast of Cape Canaveral, Florida, in 560 feet of water was determined by the President's Commission to be the part of the right booster thought to be responsible for the Challenger accident. In the area where photographs had shown a plume of flame was a one-foot by two-foot hole, lending concrete evidence to the theory that the o-ring and/or the putty used to protect it had failed to do its job of sealing two of the rocket's four segments. (*C Trib*, Apr 15/86; *W Post*, Apr 15/86; *W Times*, Apr 15/86; *NY Times*, Apr 15/86; *B Sun*, Apr 15/86)

April 17: A House panel, the Science and Technology Committee's subcommittee on space science and applications, approved a $7.6 billion NASA budget request. The authorization bill still needed review by the full Committee before being sent to the House floor. No funds were allotted for a new Shuttle, but supporters in Congress hoped that NASA would request a supplemental $500 million as first payment for "Challenger II." The physics and astronomy portion of NASA's budget, including $27.9 million for the Hubble Space Telescope, had not yet been reviewed. (*B Sun*, Apr 18/86)

April 18: An advanced reconnaissance satellite, believed to be a KH-11 photographic reconnaissance satellite, was lost as the Titan rocket carrying it exploded seconds after liftoff from Vandenberg Air Force Base in California. With only one HK-11 in orbit, officials were concerned that verification of arms control treaties would be hampered and that it would be more difficult for the Soviet Union and the United States to sign the treaties. Older satellites dropped film after photographing, but the KH-11 beamed pictures to ground stations and could maneuver about while in orbit. An HK-11 was lost in August 1985 when another Titan rocket failed. Assistant Secretary of Defense Donald Latham, estimated that it would be six or seven months before the Department of Defense could launch another satellite. Outside experts believed that the government might approve an emergency Shuttle launch from Vandenberg Air Force Base because the only other reconnaissance satellite remaining on the ground was an HK-12, believed too big for an expendable rocket. (*NY Times*, Apr 20/86; Apr 22/86; *CSM*, Apr 21/86; *USA Today*, Apr 21/86; *W Post*, Apr 22/86)

April 19: Recovery operations for the Space Shuttle Challenger crew cabin were completed. Scattered debris and the remains of all seven crew members were located in an effort that was hampered by heavy seas, high winds, and reduced underwater visibility. Final forensic work and future planning in accordance with family requests were expected to finish shortly thereafter. (NASA Release 86-46; *C Trib,* Apr 21/86; *NY Times,* Apr 21/86)

April 21: Astronaut Robert F. Overmyer announced that he planned to leave NASA, on June 1, and retire from the Marine Corps. His past accomplishments include piloting the November 1982 Columbia mission and serving as commander of the April 1985 Challenger flight. Colonel Overmyer joined NASA in September 1969. (NASA Release 86-48)

April 25: NASA announced that 25 industry/university consortiums submitted proposals for high technology research in the microgravity environment of space. Proposed areas included semiconductor crystal growth, remote sensing, communication technology, biotechnology, and space services. The proposals all have commercial potential or could contribute to possible commercial ventures. (NASA Release 86-49)

• NASA scheduled two scientific experiment rocket launchings for early May, a Black Brant X and a Taurus-Nike-Tomahawk, to create artificial clouds along the East Coast that would be visible from Canada to Florida and as far west as Ohio. The objective of the experiment was to investigate Nobel prize winner Dr. Hannes Alfven's proposed Critical Velocity Effect Theory, which has been used to explain details in the early formation of the solar system. In 1954, Alfven proposed that if an element in a nearly neutral plasma became ionized when it attained a flow velocity that matched its ionization potential, several facets of the structure of the solar system could be explained, including differing chemical compositions of the planets and the regularity of their orbits. Although Alfven's critical velocity effect has been studied in the laboratory, the phenomena required investigation in a space plasma.

The experiment was scheduled for dawn, when the Earth's surface would still be dark but when there would be sunlight at the experiment's altitude so ionized cloud chemicals could be visible to the naked eye. The three-stage, solid-propellant rockets would release barium and strontium at the apogee altitude of 267 miles. When the releases occurred, ground observers could see, if the experiment was successful, a cloud split into two well-delineated jets of gas in the first few minutes. The experiments were part of NASA's sounding rocket program, which included launching approximately 40 to 45 sounding rockets a year from various worldwide locations. (NASA Release 86-51)

• NASA disclaimed an April 23–24 two-part article in the *New York Times* that accused the organization of ignoring audits and safety standards, with particular emphasis on the Space Shuttle program. According to NASA, some of the examples mentioned in the article "are related to activities of several years ago and have since been corrected or are being corrected." Moreover, "NASA has always considered audits to be a necessary management tool to be used in conjunction with other management processes...during the existence of NASA prior to and since the Inspector General Act, management has used its own internal audits to uncover and correct deficiencies and to strengthen internal controls." In addition, NASA has been responsive to external audit reports through tighter management. "Indeed," said NASA, "$749.3 million in savings, cost avoidances, and cost recoveries were realized from 1978–1985 because of internal and external audits." NASA also noted that many of the audit reports dealt with research and development projects into which new technologies had to be figured, making for an uncertain budget environment.

Other allegations: Various violations of law and Federal codes have occurred within NASA; the Agency predicted that Shuttle cargo would cost around $100 per pound but actual cost was closer to $5264 per pound; the Shuttle was not equipped with the amount of hardware for test purposes that was available to the Apollo program; in financing a Government guaranteed loan for the Tracking and Data Relay Satellite System program (TDRSS), NASA wasted $1 billion; NASA lost $2.4 million of equipment in 1983.

In defense, NASA noted that the original estimate for Shuttle cargo was $160 per pound and the most recent was $615 per pound—besides, cost per pound was only a partial indicator of the Shuttle's utility; the Shuttle program did not parallel the Apollo program in regard to safety test hardware, but introduced longer periods of testing, not substandard safety precautions; the $1 billion said to have been wasted in the TDRSS program was in fact attributable primarily to significant interest cost increases; the lost equipment was a 1950's state-of-the-art computer that by 1983 was so obsolete that no one else would take it—hence the machine was dismantled and disposed of; ties to graft at NASA have to do with individual misconduct, such as that by Lewis Research Center and Marshall Space Flight Center officials, who were prosecuted in Federal court, indicating that the internal NASA investigation process was working well. (NASA Release 86-52)

• A statement by Shuttle Director Richard H. Truly noted that the remains of the crew of the Space Shuttle Challenger were soon to be transferred, but no positive determination of the cause of the accident had yet been made. The three approaches pursued were the examination of remains, direct examination of the wreckage, and analysis of photography and radar to determine forces imposed on the vehicle. An examination of the remains revealed no new information, and the wreckage examination was hampered by the amount of damage done at water impact. Photo and radar analysis, he hoped, would lead to a determination of the magnitude and the direction of the forces imposed on the cabin during and after breakup. (NASA Release 86-53)

April 29: NASA announced the termination of a contract with TRW to design a baseline hybrid power system for the Phase B Space Station electrical power system. TRW expressed their wish not to pursue the project, but remained a major contributor to three other work packages. (NASA Release 86-55)

• Unpublished test results done for the Presidential Commission studying the Challenger accident showed that cold weather on the morning of the Challenger launch made failure inevitable. Other findings by the commission concluded that a second joint on the right-hand booster rocket ruptured seconds after the Challenger breakup, suggesting that the joint was unduly fragile; reusable booster rockets ballooned about two-hundredths of an inch after each previous flight, and segments of the ill-fated rocket did not match well; a breach of the bottom of the external tank led to a structural failure but the nose cone of the booster rocket crashing into the fuel tank was not responsible for the final breakup; and the Shuttle did not explode, but rather a hydrogen-oxygen flash fire caused it simply to fall apart. (*NY Times,* Apr 30/86)

April 30: By a vote of 15-1, a Senate panel approved President Ronald Reagan's nomination of Dr. James C. Fletcher to head NASA. The lone negative vote came from Albert Gore, who expressed concerns about Fletcher's judgment because he had drastically underestimated the price of a Shuttle fleet in the 1970s. (*B Sun,* May 1/86; *W Times,* May 1/86; *W Post,* May 1/86; *NY Times,* May 1/86)

May

May 1: Launching of a Delta rocket carrying a weather satellite was postponed because of a quarter cup of leaking rocket fuel. The delay, said officials, would be from 1 to 10 days, depending on the extent of the problem. This was to be the first NASA launch since the Shuttle accident. (*NY Times,* May 2/86; *W Post,* May 2/86; *P Inq,* May 2/86; *B Sun,* May 2/86)

May 3: Another NASA launch ended in failure as a $30 million Delta rocket carrying a $55.7 million weather satellite was destroyed shortly after liftoff from Cape Canaveral, Florida. An order to destruct by remote control came when the rocket's main engine failed and it tumbled off course. This was the first mishap for the Delta rocket, praised as NASA's most dependable launch vehicle, in nine years and 43 consecutive launches. The four most used vehicles for launching satellites (the Delta, Titan 3D and Atlas Centaur rockets and the Space Shuttle) were all put on hold, grounding the agency for an undetermined amount of time. A short circuit was the leading suspect the Delta loss. (*W Times,* May 5/86; May 6/86; *W Post,* May 5/86; May 6/86; *WSJ,* May 5/86; May 6/86; *NY Times,* May 5/86; May 6/86)

May 6: Soviet cosmonauts made a historic trip between Space Stations as they traveled 1,800 miles in their Soyus T-15 transport module from the recently launched Mir to the Salyut-7 station. The mission was broadcast on state television. (FBIS, Tass (Eng trans), May 6/86; *B Sun,* May 7/86; *NY Times,* May 8/86)

May 9: Shuttle Director Richard H. Truly announced that John W. Thomas, Spacelab Program Officer Manager at the Marshall Space Flight Center, Huntsville, Alabama, was to assume management responsibility for the Solid Rocket Motor redesign team at the Marshall Space Flight Center on May 12, 1986. The Space Shuttle Challenger accident investigation uncovered several deficiencies in the design of the solid rocket motor joint. An independent group of senior experts would also be involved in all phases of the program; they would report to NASA Administrators and review and integrate the findings and recommendations of the Presidential Commission on the Space Shuttle Challenger Accident. (NASA Release 86-58)

May 12: Dr. James C. Fletcher became NASA's administrator for the second time. His first term in this position ran from April 1971 to May 1977. During the interim, he held the William K.Whitford Professorship of Energy Resources and Technology at the University of Pittsburgh and headed his own consulting firm. Past achievements include positions as research physicist with the

United States Navy Bureau of Ordinance, research associate at Harvard University, instructor at Princeton University, positions in the private sector with Hughes Aircraft Company, Ramo-Wooldridge Corporation and Space Electronics Corporation, and president of Space General Corporation. He also developed patents in sonar devices and missile guidance systems and has been associated with the President's Science Advisory Committee.

After being sworn in, Fletcher announced that Space Shuttle flight would not resume for another 12 to 18 months and promised to fully review NASA's quality control program and decision-making process. He also noted that the 70 percent reduction in quality control staff for which NASA was recently criticized was "misleading" because "in-house" work such as the Moon-launched Saturn rockets were no longer a responsibility. A full Senate approved his nomination on May 6. (NASA Release 86-60; *B Sun*, May 13/86; *W Post*, May 13/86; *WSJ*, May 13/86; *NY Times*, May 13/86)

• Air Force Secretary Edward Aldridge restated at a conference of the Aviation Space Writers that the USAF would like to purchase a new rocket that could lift medium weight satellites. The rockets, which the service calls "medium lift vehicles," were needed to launch 12 NAVSTAR navigation satellites scheduled for Shuttle flights. He also informed the conference of an interest in larger rockets, "complementary expendable launch vehicles," (an updated Titan rocket) for putting the Pentagon's heaviest satellites into orbit. (*B Sun*, May 14/86; *NY Times*, May 13/86; May 20/86)

May 13: NASA considered both a crew-tended station and one that would be permanently tended. The Phase B study period for the Space Station identified and evaluated alternative systems, components and philosophies that would result in a configuration responsive to the needs of potential users, be cost-effective to operate and maintain, and flexible in terms of growth, size, and capabilities. A crew-tended station could have people aboard only when a Shuttle orbiter docked to it; the orbiter would provide the habitable facilities.

While Congress directed NASA to examine an option "which phases in permanent crew features of the Station," NASA pointed out that the "phased approach would force postponement of experiments important to advancement of manned spaceflight and would increase the total cost of developing the permanently manned Space Station." NASA went on to argue that the savings achieved by the phased approach would be offset by the cost of operating the Station in a crew-tended mode for several years and by the cost of maintaining the industrial base during the delay period before resuming assembly of the permanently tended station. Savings of the crew-tended phase were estimated at $284 million in 1984 dollars and total cost prior to permanently crew-tended capability at $1 billion. (NASA Release 86-59)

• NASA Administrator Dr. James C. Fletcher appeared before a House Appropriations subcommittee and announced that he had created an independent panel to review NASA's management practices, especially the decision-making process on the morning of the January 28 Challenger accident. The review would require about six to eight months to complete. He also said that Shuttle flights would likely resume by July 1987, with an optimum of seven launches in the first year, and that funding for a replacement Shuttle was his "highest priority."

Also present was Deputy Administrator William R. Graham, who told members of Congress that NASA Inspector General Bill Colvin would investigate the transfer of two Morton Thiokol engineers, Allan McDonald and Roger Boisjoly. The two men had vehemently objected to the Challenger launch on the morning prior to its deadly flight and later detailed their scenarios before Congress. He was particularly concerned that NASA engineers might somehow be involved with the reassignments. The subcommittee advised NASA to reevaluate its relationship with the rocket booster manufacturer and threatened to severely reduce NASA's budget for new space programs if the agency did not undergo some drastic changes. (Department of Housing and Urban Development - Independent Agencies Appropriations for 1987, May 13/86; *W Times*, May 14/86; *W Post*, May 14/86; *WSJ*, May 14/86; *NY Times*, May 14/86; *P Inq*, May 14/86; *C Trib*, May 14/86; *B Sun*, May 14/86)

May 14: After more than a year of study by NASA centers and contract teams, Dr. James C. Fletcher announced a baseline configuration for the permanently crew-tended Space Station that would guide preliminary design activities for the remaining eight months of the Phase B studies. The contract called for development of the Space Station to begin in October 1986, ending with a permanently tended habitable station in 1994. A total of 14 Space Shuttle flights were thought to be required for assembly.

Major features of the design included two vertical keels 361 feet long and connected by upper and lower horizontal booms 146 feet long (the Space Station measured 503 feet at its widest point); two 44.5-foot long, 13.8-foot interior diameter U.S. supplied modules, with external interconnects for laboratory functions and crew quarters for eight members and two smaller logistics modules to alternate between the ground and the station; modules located near the station's center of gravity to provide the best possible microgravity environment for experiments, with raft patterns having four nodes and two tunnels to serve as module interconnects; internal module pressure of 14.7 per square inch and 80/20 nitrogen/oxygen mixture ratio to approximate sea level Earth atmospheric conditions, with a closed loop environmental control and life support system where oxygen and water could be recycled and nitrogen and food could be resupplied; hybrid solar power system with 75 kilowatts of power, where 25 kilowatts would be provided by photovoltaic system and the remaining 50 by the solar dynamics system; five locations on the structure for placing attached payloads and a facility for servicing free flying spacecraft and platforms with a telerobotic servicer; polar platform with useful payload on single Shuttle launch and a co-orbiting platform to support astrophysics and materials processing; gaseous hydrogen/oxygen propulsion system for altitude control (assembly would be at 220 nautical miles and operational altitude at 250 nautical miles minimum); metric as standard unit of measurement; and inclusion of international elements into the overall design, including the Canadian Mobile Servicing Center and hardware provided by Japan and the European Space Agency. (NASA Release 86-61; *NY Times,* May 15/86; *W Times,* May 15/86)

May 15: NASA and Boeing Aerospace Company entered into an agreement concerning materials processing experiments. Crystals of a size and quality impossible to create on Earth would be attempted aboard three Space Shuttle flights. Working with the University of Alabama, Boeing agreed to fund the experiments, including a chemical vapor transport crystal growth furnace installed in the galley area. These crystals were expected to be of the type valuable in the commercial production of semiconductor and electro-optic devices. The experiments would underscore NASA's commitment to the development of commercial endeavors in space, and would continue with the Space Station. (NASA Release 86-62)

• Space agency officials announced that it would cost $625 million to get the remaining Shuttle fleet off the ground. The money would go to salvage and investigation costs ($43 million), spare parts ($46 million), and corrective actions ($537 million). The corrective actions included redesign of the solid-fuel rocket booster, changes in the main engine, and changes in steering, braking and other systems, but not escape options for the crew. The Senate Appropriations Committee trimmed the request to $526 million because, it was reasoned, the agency could not spend all of its money before the fiscal year ended on September 30.

Other officials from NASA met with President Ronald Reagan to discuss the Nation's launch capability and requested $2.8 billion to replace the Challenger. The request, which was part of a $5 billion to $8 billion total that included expendable launch vehicles, violated the deficit reduction Gramm-Rudman law. Chief of Staff Donald Reagan questioned the need for a new Shuttle that might be outdated within 10 years. President Reagan queried, "Should we be buying the technology of the '70s for use in the '90s?" In grappling with the NASA budget, the President considered a recommendation to bar all commercial and foreign launches from the Shuttle. (*W Times,* May 16/86; *W Post,* May 16/86; *WSJ,* May 16/86)

May 18: An Energy Department analysis released to Representative Edward Markey stated that radioactive releases resulting from a Space Shuttle accident carrying a plutonium-powered satellite could cause 367 cancer cases and contaminate 367 square miles of land. The former estimate hinged on people staying in the launch site area for more than a year. (*W Post,* May 19/86; *C Trib,* May 19/86)

May 20: Two papers delivered by Dr. Allan Harris and Dr. Eugene Shoemaker to the American Geophysical Union in Baltimore dealt with the possibilities and consequences of a comet, asteroid, or large meteoroid striking the Earth. Chances of the first two bodies striking us in the next 100 years, the paper determined, were only 1 in 1000. An encounter with the third, however, would be more likely (this occurs about once every two decades) and although the meteoroid would explode in the atmosphere, less advanced nations might mistake the explosion for a nuclear attack. Asteroids and comets are still monitored, notwithstanding the low odds of a collision. Asteroids lend themselves to search methods because they follow an Earth-like orbit, but the latter have an elliptical orbit difficult to track.

Also presented to the Baltimore meeting were the results of the findings of a joint U.S./Brazilian atmospheric study conducted in July and August 1985 by NASA and the Instituto Nacional de Pesquisas Espaciais. Analysis of the study concluded that gases from the rain forests in the Amazon Basin in Brazil set off a chain of chemical reactions that eventually impact global air quality and the Earth's radiation budget. Trace gases in the troposphere resulting from biological activity in forest soil and vegetation were measured with instruments aboard NASA's Electra aircraft. (NASA Release 86-63; 86-64)

• The Pentagon considered waiting until 1991 before a Shuttle would be launched for the first time from Vandenberg Air Force Base in California. Temporarily closing the $2.8 billion facility would save $400 million a year in maintenance. The major advantage, however, given the increased number of launches per year, is that using only Cape Canaveral in Florida would save time because moving a Shuttle between the two coasts is time consuming. Only launches from Vandenberg, however, could safely put the Air Force's highly classified satellites into polar orbit. (*NY Times,* May 21/86)

• Representative Edward P. Bolland, chairman of the House Appropriations Subcommittee, advised NASA that it needed to make some difficult choices in its big-ticket items. The Congressman specifically asked Administrator Dr. James C. Fletcher how he would choose between a new Shuttle and the Space Station, to which he responded that the two projects were inseparable. Fletcher further pointed out that a fourth Shuttle was necessary because another accident would leave the Nation with only two, crippling the space program. The subcommittee, said Bolland, was most concerned about the cost of the Space Station. (*B Sun,* May 21/86)

May 21: Ground testing of auxiliary thrusters designed to allow the Space Station to maintain a proper orbit and altitude demonstrated that the thrusters can operate for long periods of time with no hardware degradation. Fueled by water broken down into hydrogen and oxygen, the thrusters ran for as long as 22,000 seconds, and, with a target of 10 years of Space Station life, the goal was for them to run 40,000 to 60,000 seconds. Bell Aerospace Textron and Aerojet TechSystem Corporation conducted the tests for NASA. (NASA Release 86-65)

May 22: Administrator Dr. James C. Fletcher said at a Board of Governors meeting that he expected President Ronald Reagan's approval for a replacement Shuttle. He reiterated what he told the President earlier and averred, "We need a fourth orbiter right away, as soon as we can build it." Fletcher also warned that if it were not built and if we do not begin operating the Space Station by 1994, "that could be very dire, indeed, for the future of American science and technology and the American economy." (Remarks prepared for Delivery: Aerospace Industries Association of America, Board of Governors Meeting, Williamsburg, Virginia, May 22/86; *P Inq,* May 23/86; *W Times,* May 23/86; *B Sun,* May 23/86)

May 27: Dr. James C. Fletcher said that a redesigned space vehicle (the aerospace plane) would take 10 years longer to build than a replacement for the Challenger. He also relayed the information that 44 potentially flawed parts, not including the o-rings, were discovered in the shuttle and would have to be attended to before it could again fly. (*NY Times,* May 28/86; *W Post,* May 28/86; *WSJ,* May 238/86)

May 29: NASA announced that TRW, Inc. was awarded a contract to replace the Tracking and Data Relay Satellite (TDRS) lost in the Challenger accident. Plans called for a replacement spacecraft and follow-on satellites to keep the TDRS system operational through the end of the century. As of this date, one TDRS was in geosynchronous orbit near Brazil over the Atlantic Ocean. (NASA Release 86-70; *WSJ,* May 30/86)

May 31: Another setback for Western launch capability occurred as an Ariane rocket was destroyed by ground controllers 4 minutes and 36 seconds after liftoff in French Guiana, when its third stage failed to ignite. Also lost was an Intelsat V communications satellite; it was the third time failure in the rockets had occurred for that reason. Arianespace, a French company that sold space on the rocket, was in fierce competition with NASA for multimillion-dollar commercial contracts, but ceded its monopoly when forced to postpone future flights while the accident was being investigated. With the grounding of the U.S. space program, western nations were left with the Ariane rocket as the only means to place commercial satellites into space. (*LA Herald,* Jun 1/86; *CSM,* Jun 16/86)

June

June 2: Researchers at the Construction Technology Laboratory found that concrete made from lunar soil, which contains no organic substances, was five percent stronger than high-strength Earth concrete. Because NASA planners anticipated a return to the Moon in the 21st century to establish a permanent lunar base, the concept of lunar building materials was a valuable asset when the cost of otherwise transporting them was considered. (NASA Release 86-68)

• Satellite remote sensing technology, NASA-developed software for processing satellite imagery, was used by scientists and medical researchers at NASA's National Space Technology Laboratories in conjunction with nuclear magnetic resonance (NMR) for enhancement of disease diagnosis. NMR, sensitive to soft tissue and superior to x-rays and CAT Scans, is a technology that has been around since the 1940s, but previously had not been used as a diagnostic tool. (NASA Release 86-69)

• The Presidential Commission investigating the Challenger accident released its report, two hundred pages, plus appendices, that recommended reorganization of NASA's management structure. Other changes cited by the Commission included more rigid quality control, reorganization of the lines of communication in the NASA chain of command, a larger role for NASA astronauts in Shuttle matters, and a Shuttle escape system. By the time these conclusions were announced, the report had been sent to the printers; it was scheduled to be on President Ronald Reagan's desk by June 9. (Report of the Presidential Commission on the Space Shuttle Challenger Accident; *W Post,* Jun 3/86; Jun 10/86; *B Sun,* Jun 3/86; Jun 10/86; *NY Times,* Jun 4/86; Jun 10/86; *WSJ,* Jun 4/86; Jun 10/86)

June 3: NASA announced the seven winners of the sixth annual Space Science Student Involvement Program (SSIP) and the winner of the SSIP Student Newspaper Competition. SSIP, a joint venture of NASA and the National Science Teachers Association, and the corresponding newspaper competition were both aimed at the secondary school level. (NASA Release 86-71)

• Preliminary investigation into the Titan rocket lost on April 18 found that the insulation between its solid-fuel and the outer casing of the solid rocket booster might have peeled, allowing flames to burn through its casing. Although made by United Technologies, the design was identical to that used in the boosters made by Morton Thiokol for the Space Shuttles and raised the question as to whether solid-fuel booster rockets—in lieu of liquid-propellant rockets were suitable for human space flight. (NY Times, June 4/86)

June 4: NASA located a video tape concerning the Shuttle mission 51-E Flight Readiness Review (FRR) that contained two-and-one-half minutes of discussion on previous o-ring erosion. Because there was no requirement to record audio or video of FRR's, no cataloging of the tape was made. The Presidential Commission was provided with the tape but informed NASA that no new information was on it. (NASA 86-77)

June 5: Two near-Earth asteroids, composed of nickel and iron and could provide future space mining, were discovered shortly before this date by NASA's Jet Propulsion Laboratory. About one mile in diameter, each, one asteroid is inside and one is outside of Earth's orbit. (NASA Release 86-73)

• Two hundred fifty-eight members of Congress signed a letter addressed to NASA urging them to find a second supplier for its Space Shuttle booster rockets. Congress, described as further tightening the rein on NASA, believed that competition for the sole provider, Morton Thiokol, would equate to savings. Aerojet General said it was ready to build a seamless booster, a safer design. Congress also wanted NASA to eliminate some of the rules that favored Thiokol, such as requiring any second manufacturer to purchase nozzles for the rocket from them. (*W Times,* Jun 6/86; *WSJ,* Jun 6/86; *P Inq,* Jun 6/86; *B Sun,* Jun 6/86)

• Investigation into the Delta rocket crash May 3 indicated that the accident could have been a result of a faulty hardware-assembly program in the engine compartment. A short circuit was most likely the cause, said Lawrence Ross, Chairman of the Delta review board, but sabotage was not ruled out. (*B Sun,* Jun 6/86)

June 9: Following the release of the Presidential Commission's report, NASA Administrator Dr. James C. Fletcher announced that he would adopt most of the recommendations. He stated, "The whole agency needs a major reexamination." He further said that he still expected the Shuttle fleet to resume operation by July 1987, that the President will opt for a replacement Shuttle, and that he remained committed to the Teacher in Space Program. Congress, however, pointed out that fiscal restraints would render it nearly impossible to make all the Committee's suggested repairs while building a new orbiter and pursuing the Space Station and the aerospace plane. (NASA Release: Statement of Dr. James C. Fletcher, NASA Administrator, Jun 9/86; *W Post,* Jun 10/86; *NY Times,* Jun 10/86)

June 10: Hearings before the House of Representatives investigating the Challenger accident began on this date. Testimonies were given by William P. Rogers, Dr. James C. Fletcher, various NASA personnel, various personnel from Morton Thiokol, Inc., and various personnel from the military and industry. The hearings were to take place June 10, 11, 12, 17, 18, and 25, 1986 and July 15, 16, 23, and 24, 1986. (Investigation into the Challenger Accident: Hearings before the Committee on Science and Technology House of Representatives, 1986)

June 11: NASA installed a super computer known as the Cray X-MP for use in key NASA research projects and for sharing with other NASA Centers, industries, and universities. The computer made possible the analysis of problems previously thought impossible to solve. A solution that might take one hour on the Cray X-MP, for example, would take 200 hours on a popular business mainframe computer. NASA noted that the research underway at Lewis's Structural Mechanics Branch would have been prohibitively expensive and time consuming without the super computer. (NASA Release 86-74)

• NASA officials said that resuming Space Shuttle flight by July 1987 would be impossible if all of the conclusions reached by the Presidential Commission were followed. The report called for correction of the Shuttle's braking system, development of an escape system, and review of more than 700 critical components; however, the biggest obstacle lay in the redesigning and testing of the booster rockets. The Commission recommended a costly and time consuming full-scale, vertical test for the new rockets. NASA had relied on a less expensive, scaled down, horizontal test for new designs in the past and made no commitment for following this advice. The added preparation time would most adversely affect the Air Force, who earlier had warned that even a 1-year delay would produce a backlog of 40 payloads by 1992; some of their satellites could not fit aboard any untended launch vehicle then in use. Some former commercial customers had negotiated flights with the French Ariane rocket program, but because they were also grounded, China and Japan were being sought for launches.

President Ronald Reagan, in the meantime, had called for a replacement orbiter, but had not yet revealed how it would be funded. (*NY Times,* Jun 12/86)

• Astronaut Owen K. Garriott planned to leave both NASA and government service during the current week. His past achievements include a 2-month stay in space aboard Skylab and the 10-day Spacelab 1 mission aboard the Space Shuttle Columbia. (NASA Release 86-75)

June 12: Senior officials at the White House found it increasingly difficult to go ahead with President Ronald Reagan's call for another Space Shuttle. The President's remarks, they inferred, did not constitute any commitment.

Meanwhile, members of the House committee, delving further into the Challenger accident, questioned Shuttle Director Richard H. Truly in regards to the Marshall Space Flight Center (MSFC), Huntsville, Alabama, which was in charge of the booster rocket. Representative James H. Scheuer believed that information regarding the faulty booster was known not only at the MSFC but by officials at NASA Headquarters. (*W Post*, June 13/86; *WSJ*, June 13/86)

• The Soviet Union said that it was prepared to fill the void left by the United States and Europe for commercial space launches. Moscow revealed this fact in an announcement that also called for a World Space Organization. The Organization would oversee joint space projects that the Soviets hoped would unfold in the future. (*C Trib*, Jun 13/86; *CSM*, Jun 16/86)

June 13: In a move that would greatly reduce the chances for resuming Shuttle flights by July 1987, President Ronald Reagan urged NASA Administrator Dr. James C. Fletcher to comply with all of the recommendations put forth by the Challenger commission. President Reagan also ordered Fletcher to report back to him after 30 days with details on how and when the findings would be implemented. Also, the two reportedly discussed approval of Federal funds for a replacement orbiter and the President's final decision was expected to follow after a few days. (Letter from President Reagan to James Fletcher, Jun 13/86; *W Post*, Jun 14/86; *WSJ*, Jun 16/86; *W Times*, Jun 16/86)

June 15: The Peoples Republic of China moved to fill the void left by the United States and Europe for commercial space launchings as it negotiated with two U.S. firms. Hughes Aircraft Company expressed an interest in working with China to establish a launch base in the Hawaiian islands. Peking, at the same time, announced an agreement it had made with the U.S. company Teresat, Inc., for putting two of their satellites into orbit in 1987 and 1989 at a price ranging from $20 million to $25 million, about 15 percent less than what Ariane and the Space Shuttle charged. Teresat, however, would first need to locate an insurer for launches and secure approval from the State Department. China claimed its program was capable of 10 to 12 launches per year. (*W Post*, Jun 16/86; *CSM*, Jun 16/86; *WSJ*, Jun 17/86; *C Trib*, Jun 17/86)

June 17: Astronaut James Van Hoften announced his resignation from NASA to enter private industry. His past achievements include two Space Shuttle Flights and the position of Assistant Professor of Civil Engineering at the University of Houston. He was the fifth astronaut to leave NASA since the January 28 accident. (NASA Release 86-78; *USA Today*, Jun 18/86; *LA Times*, Jul 29/86)

• The $3 billion launch site at Vandenberg Air Force Base in California was deemed unsafe according to a Senate report released on this date. The salient feature cited was the closed exhaust ducts that could fill with hydrogen gas in the event that a Shuttle's main engines were shut down to abort liftoff or at the end of a flight-readiness test-firing. The report also noted that fog, high winds, and cool temperatures were common to the area and that the proximity of the launch pad to the control center put support personnel at risk. Because the Department of Defense required the California site to achieve polar orbit for reconnaissance satellites, it was recommended that they adopt the "new" Titan missiles. (*W Times,* Jun 18/86; *LA Times,* Jun 18/86; Jul 9/86)

June 18: Allan McDonald, in charge of the booster rocket redesign task force, told members of Congress that the new booster joints would fall into the "criticality-1R" category, meaning that they would have a backup. No specific design had yet been chosen, and the engineer stated, "We're looking at some designs with o-rings and some without. We're looking at various seating concepts." McDonald was taken off Morton Thiokol's solid rocket team after testifying before the Rogers Commission concerning his objection to launching the Challenger in cold weather, but was reassigned to head the task force after complaints from Congress. (USA Today, Jun 19/86; W Times, Jun 19/86; NY Times, Jun 19/86)

June 19: Lawrence F. Herbolsheimer was appointed Deputy Assistant Administrator for the Office of Commercial Programs, where he would be responsible for advancing the interests and participation of the private sector in the U.S. Space Program. He previously had co-founded businesses in the private sector and had served at the White House. (NASA Release 86-79)

• NASA announced that it would terminate further development of the $1 billion Centaur Upper Stage for launching other spacecraft on board a Space Shuttle. The Shuttle 51-L accident prompted major safety reviews and safety criteria, which the Centaur, even after modifications, could not meet. Under Richard H. Truly's direction, NASA initiated efforts to examine alternatives for those national security missions that had planned to use Centaur. Because the Ulysses and Galileo missions were dependant on such a launch, other options for launching them from the Shuttle with a Titan rocket were being considered. (NASA Release 86-80, *W Times,* Jun 20/86; *W Post,* Jun 20/86; *WSJ,* Jun 20/86; *NY Times,* Jun 20/86; *CSM,* Jun 23/86)

June 24: Following a speech in Washington, Air Force Secretary Edward Aldridge told reporters that the Space Shuttle would not fly again until early 1988. The Air Force, he said, would use the expendable launch vehicles being developed as often as possible and that NASA's goal was to have military and non-military payloads on a 50-50 schedule. For reasons of security, Aldridge had hoped President Ronald Reagan would authorize funds for a new orbiter, but sources revealed that the President would hold back any decisions concerning the space program until after NASA began implementing some of the recommendations from the Rogers Commission. (*W Post,* Jun 24/86; *W Times,* Jun 24/86; *NY Times,* Jun 24/86)

June 25: TRW, Inc. was selected by NASA for negotiations leading to the award of a contract for design and delivery of the Orbital Maneuvering Vehicle (OMV), a reusable, remotely operated propulsion vehicle for hauling satellites from one orbit to another. The OMV's primary uses would be spacecraft delivery, retrieval, boost, deboost, and close proximity visual observation beyond the operating range of a Space Shuttle. The OMV, "space tug," was also to be made to adapt for use in Space Station activities. (NASA Release 86-81; *WSJ,* Jun 26/86; *LA Times,* Jun 27/86)

• The first flight of an aircraft with both a digital jet engine control system and a mated digital flight control system took place at NASA's Ames-Dryden Flight Research Facility with a modified F-15. The integration of engine and flight control (Highly Integrated Digital Electronic Control) between the system's computers could increase performance 2 to 10 percent. (NASA Release 86-82)

June 26: NASA officials confirmed that the Agency had considered reproduction of the Delta rocket and was ordering at least one from McDonnell Douglas for launching an Indonesian communications satellite early in 1987. Officials at the corporation assured NASA that the grounded Delta rocket could be ready for launch in a few months because problems were worked out and the production line for the rocket was partially restarted. (*W Post,* Jun 27/86; *WSJ,* Jun 27/86)

• Three government agencies—the National Oceanic and Atmospheric Administration (NOAA), the National Science Foundation (NSF), and the National Aeronautics and Space Administration (NASA)—proposed a long-term plan for the study of human effect on the Earth's environment. The project envisioned cooperation from other scientifically advanced nations and would use the disciplines of geology, meteorology, oceanography, forestry, agriculture, and computer and satellite technologies to study the Earth as an integrated system. NSF director, Erich Bloch, noted that "for the first time in history we have the capability to observe the entire Earth from the outer reaches of its atmosphere to its molten inner core." NASA's Earth System Sciences Committee predicted that the program, enlisting satellites from countries around the globe, would be in full swing by the middle 1990s. (*B Sun,* Jun 27/86; *CSM,* Jun 30/86)

• For the first time since the January 28 accident, the Space Shuttle's main engine underwent a 1-1/2 second test for ignition. It was an uneventful, successful run, and NASA officials offered it as a signal that the program was gearing up for a targeted July 1987 mission. (*B Sun,* Jun 27/86; *C Trib,* Jun 27/86)

June 30: Dr. James C. Fletcher, NASA Administrator, announced that Andrew J. Stofan had been appointed Associate Administrator for the Space Station. He also announced a number of organizational and management structural changes for the program and transfer of its supervision from the Johnson

Space Center, Houston, Texas, to Washington, D.C. The changes followed recommendations from a committee headed by former Apollo Program Manager Samuel C. Phillips that were consistent with findings from the Rogers Commission, which investigated the Challenger accident. Phillips examined the Space Station program from both a technical and management perspective and Dr. Fletcher, in turn, made changes that assured the flow of important information to proper decision-making levels. Another result was the procurement of a versatile computer-based information network to link NASA and contractor facilities and to provide engineering services. (NASA Release 86-84; *W Post*, Jul 1/86; *NY Times*, Jul 1/86; *W Times*, Jul 1/86; *C Trib*, Jul 2/86)

July

July 1: Al Louviere, who headed a team of engineers at the Johnson Space Center, Houston, Texas, recommended to Shuttle Program Director Arnold Aldrich that an escape system be built into orbiters. Equipped with pressure suits and an oxygen supply, astronauts could bail out of the aircraft while it was in gliding flight below 100,000 feet. Although no specific plan was given, Louviere stressed the fact that no escape system could have saved the Challenger crew. (*P Inq,* Jul 2/86)

July 2: The Delta 178 rocket accident that occurred on May 3, 1986, was caused by mechanical damage to wiring caused by vibration during flight. The rocket's first failure—after 43 flights since 1977—was attributed to a design change, a shift from polyvinylchloride (PVC) insulated wiring to Teflon insulated. Unlike PVC, which is over-wrapped for greater abrasion protection, Teflon wiring bundles are over-wrapped in only a very limited number of locations; the abrasion resistance of mechanical damping was not adequately considered when making the change. The investigating board recommended a review of the booster electrical control system and possible design changes, verification of the quality of all connectors, and a reemphasis on highly attentive, quality workmanship.

Another study released concerned the Titan 34-D mishap of April 18; it determined the cause of the accident to be insulation inside one of the two solid rockets being pulled away and allowing "burn through". Neither accident was deemed the result of any major design flaw and, subsequent to minor modifications, the United States planned to resume untended rocket launches.

The investigating board also observed potential problems resulting from high-humidity conditions for vehicles stored at the Kennedy Space Center, to instrument limitations, and lack of protection from contamination of the Rocketdyne engine relay box. These problems were not directly connected with the Delta 178 rocket accident or the Titan 34-D mishap. (NASA Release 86-85; *W Post,* Jul 3/86; *WSJ,* Jul 3/86; *NY Times,* Jul 3/86; *B Sun;* Jul 3/86)

July 3: NASA issued a request for proposals to various contractors for a satellite for use in a joint venture with the French Space Agency, Centre National d'Etudes Spatiales, in the Topex/Poseidon oceanographic mission. The satellite would map the circulation of the world's oceans by using a precise radar altimeter to measure height variations on the sea surface. Contingent upon congressional approval of funds, a launch of the satellite in late 1991 was planned. (NASA Release 86-86)

July 7: A full-scale Spacehab module mockup intended to double the pressurized volume of the Space Shuttle mid-deck was delivered to NASA's Ames Research Center by Spacehab, Inc. The mockup was intended as a test bed for artificial intelligence, robotics systems, and human performance. Spacehab planned to build three of the modules over the next three years.

Another corporation, Sverdrup Technology, was chosen for a $45 million support services contract in which they would provide test complex services for Space Shuttle main engines and engineering and laboratory and data work for NASA and 18 other Government agencies. NASA also planned a program that would develop software tools for writing computer programs in ADA, a language used by the Department of Defense for mission-critical computers. The project was expected to run for three or four years and cost $200 million. (*Def News,* Jul 7/86)

July 8: In a response to the Presidential Commission on the Space Shuttle Challenger Accident, Dr. James C. Fletcher announced the establishment of the Office of Safety, Reliability and Quality Assurance and the appointment of George A. Rodney, from Martin Marietta Corporation, to head the office as Associate Administrator. Objectives of the office were to ensure a program that monitored equipment status and designed validation problem analysis and system acceptability in Agencywide plans and programs. Rodney's responsibilities were to have oversight of safety, reliability, and quality assurance functions related to all NASA activities. (NASA Release 86-87; *W Post,* Jul 9/86; *W Times,* Jul 9/86; *B Sun,* July 9/86; *P Inq,* Jul 6/86)

July 10: With a request for proposals issued to approximately 250 firms, NASA planned to implement a Technical and Management Information System (TMIS) to support overall functions of the Space Station, including design, development, and operation. The TMIS would link the various NASA centers involved with the program and would automate the generation and interchange of documents, correspondence, schedules, engineering data and drawing, budget data, and other management information. Carried out in implements, the first would be operational less than one year after the contract award date. (NASA Release 86-88)

• Richard G. Smith, Director of the Kennedy Space Center in Florida, announced his retirement. He plans to join the private sector with General Space Corporation. His achievements span 35 years in the space program and include work with the rocket research and development team at Redstone Arsenal; management of the Saturn Rocket program at Marshall Space Flight Center, Deputy Director of the Marshall Space Flight Center, Deputy Associate Administrator for Space Transportation Systems at NASA Headquarters, Director of the Skylab Task Force, and he is the recipient of awards for his contributions to Apollo, Skylab, and Space Shuttle programs.

Smith had earlier warned that investigation into the Challenger accident would result in a mass exodus from NASA and later added that the investigation damaged the reputations of many officials. (NASA Release 86-89; *W Times*, Jul 11/86; *W Post*, Jul 11/86; *NY Times*, Jul 11/86; *LA Times*, Jul 11/86)

July 14: NASA planned for a display of current and future technologies at the 34th Annual EAA International Fly-In Convention and Sport Aviation Exhibition, with an exhibit entitled "21st Century Aviation," August 1–8 in Oshkosh, Wisconsin. Featured would be hypersonic cruise vehicles and high altitude aircraft that could fly from the U.S. West Coast to Tokyo five times faster than subsonic flyers and have the capability of taking payloads and passengers to Earth orbit. In addition, NASA also planned advanced cockpit displays in which aircraft controls reacted to voice commands, and presentations of a supercomputer capable of one billion computations per second, rotocraft research and potential applications, aeronautical art, and the Search and Rescue Satellite (SARSAT) system. The latter display was intended to show how using a satellite saves time in rescuing air and maritime casualties. (NASA Release 86-90)

• In response to a recent order, NASA Administrator Dr. James C. Fletcher handed President Ronald Reagan a 50-page report that outlined the agency's plans for implementing the Rogers Commission's recommendations. Among the items detailed were booster redesign, changes in NASA's management, review of "critical items" whose failure would lead to vehicle loss, establishment of a safety organization, improved communications, landing safety, launch abort and crew escape, flight rate, and maintenance safeguards. Fletcher also submitted a companion letter urging the President to authorize funds for replacement of the Challenger. Because of the requirements for getting the Shuttle flying again, NASA confirmed the aircraft would be grounded until at least early 1988. (Actions to Implement the Recommendations of the Presidential Commission on the Space Shuttle Challenger Accident, Jul 14/86; *NY Times*, Jul 15/86; *W Post*, Jul 15/86; *B Sun*, Jul 15/86; *W Times*, Jul 15/86; *P Inq*, Jul 15/86; *C Trib*, Jul 15/86)

July 16: Lawrence B. Mulloy, a manager at the Marshall Space Flight Center, Huntsville, Alabama, in charge of the solid rocket motors that were held responsible for the Challenger accident, announced his retirement after more than 30 years of Government service. He was among those who opted for liftoff of the ill-fated Shuttle and was repeatedly faulted for this judgement by William P. Rogers, chairman of the commission investigating the accident. He was the third Marshall manager to retire since the accident, following William R. Lucas and George B. Hardy. (*W Post*, Jul 17/86; *P Inq*, Jul 17/86; *B Sun*, Jul 17/86)

July 17: NASA got back tapes from IBM that were recovered from the Atlantic Ocean six weeks after the Space Shuttle accident and revealed that the Challenger crew had no knowledge of events leading to the catastrophe. IBM engineers, who were simultaneously researching tape restoration, were called upon and after isolating the computer zeros and ones, NASA engineers began the job of translating the data into a facsimile of human voice. (*W Post,* Jul 18/86; *NY Times,* Jul 18/86; *B Sun,* Jul 18/86)

July 18: Shuttle director Richard H. Truly announced NASA's interest in an improved, second-generation solid rocket motor design for study by the National Space Transportation System program manager. Among the requested changes were asbestos-free insulation, alternate case and propellant design, and modified burn rate for improved performance. Outside geometry and interfaces with other Space Shuttle elements would not be affected. He added, however, that the tentative schedule for launching a Shuttle in early 1988 would not be contingent upon this newly designed booster rocket. (NASA Release 86-94; *NY Times,* Jul 21/86)

July 21: Conrail, created by Congress from six bankrupt northeastern railroads, argued against a measure proposed by Senator John Danforth that would take $450 million from the Government-owned railroad and transfer the funds to NASA. If passed, the plan would help compensate for the House and Senate Armed Services Committee's move to cut $566 million from the Air Force budget that was earmarked for NASA and military use of the Shuttle. (*W Post,* Jul 22/86)

July 22: Testing of recently installed weather protection structures was scheduled for August 19 at the Kennedy Space Center in Florida with Space Shuttle Atlantis. The $3.3 million system of sliding and folding doors and seals would cover previously exposed, lower portions of the orbiter and greatly reduce risk to the heat protection tiles, subject to damage from heavy rains, hail, and wind-blown debris. (NASA Release 86-95)

• Thomas Paine, Laurel Wilkening, George Field, Jack Kerrebrock, Kathryn Sullivan, and Carl Sagan testified before Congress concerning Paine's report, "Pioneering the Space Frontier." (Hearing before the Committee on Science and Technology, July 22/86)

July 23: The National Scientific Balloon Facility found a promising new material in a flight test series of high altitude balloons. The balloon film, developed by Raven Industries, was 0.0008 inches thick; two successful flights carried one 3,000-pound payload to an altitude of 130,000 feet and another 4,300-pound payload to 120,000 feet. At that time, the NASA balloon program provided support for approximately 50 flights-per-year for studies primarily in high energy astrophysics, astronomy, and upper atmospheric research. Because of its weight-carrying capacity and flight duration, the balloon was also important in infrared astronomy research and in cosmic and gamma ray investigations. (NASA Release 86-96)

• NASA announced another test of a Space Shuttle engine to take place later in that week. The purpose was to collect additional data to be added to the previous test of the same engine, that ran for 250 seconds, at the National Space Technology Laboratory. (NASA Release 86-97)

• NASA officials announced that the launch of an Atlas-Centaur rocket scheduled to carry a Navy communications satellite into orbit on August 28 was postponed indefinitely. The flight was originally set for May 22, but the Delta rocket accident canceled that liftoff. The Atlas was the only remaining U.S. rocket system that had not been grounded; the delay resulted from the discovery by engineers of problems with some of the rocket's electrical parts. (*W Post,* Jul 24/86)

July 24: Dr. James C. Fletcher released to members of Congress a statement indicating that employment at the Johnson Space Center (JSC), Houston, Texas, would increase from 12,600 to 14,000 employees during Space Station development, with a potential for another 1,000 jobs should JSC take on additional Space Station tasks. "Another way to look at this," he warned "is that without the Space Station program, JSC's employment level likely will not grow significantly over the next few years."

The projected increase notwithstanding, officials at the JSC were concerned with their status because of the recent decision to move the habitation module's development to the Marshall Space Flight Center, Huntsville Alabama. They were assured, however, that the move had to do with coordinating the habitation module and the module structure and that "the Johnson Space Center's role as NASA's preeminent center of excellence in manned spacecraft systems is not changing." (NASA Release 86-98; *NY Times,* Jul 26/86; *W Post,* Jul 26/86)

July 25: Because of NASA's expertise in microelectronics and other miniaturized instruments, its technology was studied for use in controlling hydrocephalus, an impairment characterized by an accumulation of fluids in the brain that can lead to mental disorders. The microprocessor-controlled cerebrospinal fluid outflow regulating system would replace an implanted shunt that diverts fluid from the brain to other parts of the body. Because the shunt is not always successful and often needs repair, this new technology, which can be reprogrammed to meet changing needs, offered hope for improvement and stability. (NASA Release 86-99)

July 28: Following a review of the wreckage and all other available data from the Challenger flight, copies of the final report on the investigation into the cause of death of the crew were released. Although NASA was unable to arrive at a positive determination, it seemed possible, after examining the transcripts from internal communication, that the crew was unaware of the events immediately preceding the breakup of the orbiter.

Release of the Challenger tapes to the news media, however, revealed that some members of the crew may have been aware of the accident. Seconds before the crash, pilot Michael Smith uttered the phrase "Oh-oh" and subsequent to that other evidence revealed that a few astronauts attempted to use emergency air supply. The exact cause of death was never determined, but sources agreed that it was doubtful that the initial blast was sufficient to cause death or injury and whether any of the crew were able to maintain consciousness until the cabin struck the ocean's surface could not be determined. (NASA Release 86-100; *W Post,* Jul 29/86; *W Times,* Jul 29/86; *NY Times,* Jul 29/86; *WSJ,* Aug 5/86)

July 29: Administration sources said that President Ronald Reagan might delay funding for a new orbiter until 1987. Because budget restraints made any commitment to a replacement impossible, many advisors were in favor of producing expendable launch vehicles for private satellites. (W Post, Jul 30/86; P Inq, Jul 30/86; WSJ, Jul 31/86)

July 31: NASA announced four teams to establish centers for the Commercial Development of Space. The centers, joint undertakings of academic, industry, and government teams, were (1) University of Wisconsin, Space Automation and Robotics and Dr. John Bollinger; (2) Ohio State University, Real Time Satellite Mapping and Ivan Mueller; (3) University of Houston, Molecular Beam Epitaxy and C.W. Chu; and (4) Clarkson University, Commercial Crystal Growth in Space and William Wilcox. Criteria for selection were based on new and unique research leading to commercial activity, management teams capable of directing commercial research, and available non-NASA resources to help operate the center. (NASA Release 86-102)

• Dr. James C. Fletcher acknowledged that NASA did not sufficiently consult Congress on decisions pertaining to Space Station management prior to the June 30 announcement and said he would postpone plans for moving Space Station management from Houston to the Washington, D.C. headquarters. Because "considerable misunderstanding and misinterpretation of the intent of the decision on the Space Station work package realignment has resulted," said Fletcher "I intend to conduct a thorough review of all aspects of Space Station design, work package assignments and functions, and conduct extensive conversations with members of Congress." (NASA Release 86-103; *USA Today,* Aug 1/86)

• White House officials announced that top members of President Ronald Reagan's cabinet were in favor of halting commercial launches aboard the Space Shuttle, a move that would cost NASA $850 million in revenue. White House spokesman Larry Speaks argued that eliminating the launches would stimulate the private aerospace industry, reluctant to compete with NASA in the past, to quickly fill the void. Another reason for the reduced number of launches, he noted, was the increasing backlog of scientific and military payloads. Focus on those two types of missions exclusively would ease the immediate need for a fourth shuttle.

At the same time, Air Force Secretary Edward Aldridge said that he intended to ask Congress for $2.6 billion in order to acquire a fleet of rockets that would reduce both the backlog of 21 payloads and the Pentagon's dependance on the Space Shuttle. He added that no Space Shuttles would be launched from California's Vandenberg Air Force Base until 1992, a saving to tax payers of $1 billion. (*W Times*, Aug 1/86; *WSJ*, Aug 1/86; *W Post*, Aug 1/86; *B Sun*, Aug 1/86; *NY Times*, Aug 1/86; *WSJ*, Aug 5/86)

August

August 1: The Lunar and Planetary Laboratory at the University of Arizona and NASA's Johnson Space Center in Houston, Texas, planned a Planet Mercury Conference for early August. More than 10 years had elapsed since the last conference and a significant amount of data had developed during the interim. Scheduled topics included Mercury's role in helping scientists understand the origin of planets, Mercury as an end-member planet, ground-based and Earth-orbital observations of Mercury, and possible spacecraft missions to the planet. (NASA Release)

• In Space Station Phase B Program Level Agreement, NASA and the ESA announced agreement on hardware elements for a preliminary design of a permanently attached pressurized laboratory module and a polar orbiting platform, the latter for conducting Earth observations. Under the agreement, ESA would conduct research on a preliminary design of a man-tended free-flyer (pressurized module and resource module) for international utilization, primarily in the fields of material and life sciences and fluid physics, that would require both a long-duration and undisturbed microgravity environment. (NASA Release 86-104)

August 5: Preliminary data from NASA's Earth Radiation Budget Experiment (ERBE), a three-satellite project that began in October 1984, indicated that clouds reflect more heat than they retain and appear to cool Earth's climate, possibly offsetting the "greenhouse effect". Incomplete coverage and sporadic observations have limited decades of study on global heat flows done with sounding rockets, balloons, and satellites. In a broader attempt, the ERBE instruments measured the average monthly heat budget on regional, zonal, and global scales; tracked the seasonal movement of heat from the tropics to the poles; and determined the average daily variation in heat on a 620-mile regional monthly scale. An ERBE package contained two radiometer instruments. One, a scanner, was used for narrow field-of-view scanning with short-wave measurements of reflected solar energy and long-wave measurements of Earth-emitted energy. The other, a non-scanner, was used for wide field-of-view of the entire disc of the Earth, a 10-degree region of Earth, and the total output of the sun's radiant energy. (NASA Release 86-105)

• NASA announced that James R. Thompson would be appointed Director of the Marshall Space Flight Center in Huntsville, Alabama, in late September. He had been Deputy Director of the Princeton Plasma Physics Laboratory since 1983 and previous to that had spent 23 years at Marshall, where he managed development of the Space Shuttle's main engines, worked in the Skylab

Program, and was Associate Director of Engineering for the Center. He was also NASA's manager for the Challenger investigation (NASA Release 86-106; *NY Times,* Aug 6/86)

• Dr. James C. Fletcher appeared before a Senate subcommittee and testified that the $7.7 billion budget for NASA was "barely adequate" and could further delay Shuttle flights while possibly stripping the private sector of some 25,000 jobs. He also said that the currently designed Space Station was intended to be built using four Shuttles. With only three Shuttles, the Station would need to go back to the drawing board. The Senate appeared to be looking at a civilian sponsored space program for commercial launches as a way to reduce costs. (*W Post,* Aug 6/86; *B Sun,* Aug 6/86)

• Documents released by the Rogers Commission investigating the Challenger accident revealed that the Space Shuttle Columbia had narrowly avoided a severe mishap on January 6 when it came within 31 seconds of launch after 18,000 pounds of liquid oxygen had been inadvertently drained from its external tank. The mistake would have left the ship without enough fuel to obtain orbit; the error was blamed on fatigue among the ground crew, who were pursuing a rigid schedule. (*NY Times,* Aug 6/86; *W Post,* Aug 6/86; *W Times,* Aug 6/86; *B Sun,* Aug 6/86)

August 7: NASA and the Space Telescope Science Institute announced that U.S. amateur astronomers would be given a small amount of observing time with the Hubble Space Telescope (HST). Amateur astronomers have made significant contributions to astronomy, and, at this point, there are at least a quarter million of these astronomers. With a 94-1/2-inch mirror, the HST would be the largest astronomical telescope ever placed in space; it serves as an example of international cooperation between NASA and the European Space Agency. (NASA Release 86-108)

• In a White House meeting President Ronald Reagan indicated that he would allow NASA to honor only some of its higher priority contracts for commercial and foreign satellite launches from the Space Shuttle. Regarding a new orbiter, Chief of Staff Donald Regan has reversed his stance and now favors the replacement. Regan changed his mind, he stated, because he believed that NASA would not just correct its major faults, but would bring the Shuttle up to date with current technology. In a related matter, the Senate agreed to reimburse NASA with $56.3 million for military launches, a move protested by Senator Sam Nunn, who claimed that the money would come at the expense of intelligence activities. (*NY Times,* Aug 8/86; *W Post,* Aug 8/86)

August 11: As of this date, two companies (Boeing Aerospace Operations and McDonnell Douglas Astronautics Company) had submitted proposals for the

Payload Ground Operations Contract, a three-year cost-plus-award-fee contract with optional extensions of up to 15 years. The selected contractor would be responsible for preparing all payload for launch aboard Space Shuttles and expendable launch vehicles, as well as operations, maintenance, and engineering of the related processing facilities. The contract would consolidate work that was performed under separate, on-site contracts involving about 900 people. Space Station components requiring launch from the Kennedy Space Center and preparation of NASA payloads at the Vandenberg Launch Site in California were to be included. (NASA Release 86-109)

August 12: NASA chose a new design for its solid rocket motors, a major step for returning the Shuttle fleet to flight. Each joint would be sealed by three o-rings constructed from a more resilient material than that used originally, and would seal at all times, not just from pressure produced by expanding gas inside the rocket. Other changes included a metal clip meant to force the joint to remain sealed, interlocking insulation to prevent hot gases from directly contacting the o-rings, and a bolt assembly between the rocket nozzle and the engine for providing an additional seal at the nozzle joint. Hundreds of tests for the new design were scheduled for the coming fall. (*NY Times,* Aug 12/86; Aug 13/86; *W Times,* Aug 12/86; *W Post,* Aug 13/86; *C Trib,* Aug 13/86)

August 13: Pratt & Whitney was selected by NASA for negotiations leading to a contract award for the design, development, test, flight certification, and production verification of alternate high-pressure fuel and oxidizer turbopumps. With a proposed cost of about $182 million and a period of performance of five to six years, the turbopumps were intended to be interchangeable with the Space Shuttle's main engine turbopumps, provide extended life capability, and enhance safety margins. (NASA Release 86-112)

August 14: Members of a House Space Subcommittee endorsed NASA's $7.7 billion budget and authorized funds for a new shuttle and President Ronald Reagan ordered that a replacement orbiter be built. The craft would take about seven years to construct, according to administration officials. The money for the project, earmarked to come from other space programs, would be allotted annually. The first payment was to be $200 million, and the rate of progress for building it depended on future financing. The President also restated his hopes for private industry to develop a capacity for launching commercial satellites. (*W Times,* Aug 14/86; Aug 15/86; *B Sun,* Aug 14/86; Aug 15/86; *NY Times,* Aug 15/86; *C Trib,* Aug 15/86; *W Post,* Aug 15/86; *CSM,* Aug 18/86)

August 15: In a statement released by the White House Press Secretary, President Ronald Reagan announced "two steps that will ensure America's leadership in space exploration and utilization." His first target was the building of a fourth Space Shuttle beginning in fiscal year 1987, deemed necessary for accomplishments in space (especially the Space Station). He also announced that NASA would ease out of the business of launching commercial satellites by allowing the private sector to assume the responsibility, which, he said, would eventually be done "better and cheaper." NASA, he stressed, should focus solely on the exploration of space. (The White House, Office of the Press Secretary, Statement by the President, Aug 15/86)

August 18: A Gamma Ray Imaging Telescope for studying gamma ray sources in the universe was the basis for a $93,000 contract awarded to Martin Marietta. By housing the telescope inside of a Space Shuttle external liquid hydrogen tank, which is normally discarded just before the Shuttle achieves orbit, Martin Marietta deemed it possible that the unit could also be set into orbit. Telescope components would be carried in the cargo bay and assembled inside the emptied tank by astronauts, who could enter via an existing 36-inch aft manhole port or through tank modifications. The satellite could then be pressurized to provide the needed environment for gamma ray detection technique; the rays would be converted to positrons and electrons that travel the length of the telescope emitting light. Gamma ray astronomy is essential to understanding the evolution of stars and the universe and the physical processes occurring in pulsars, quasars, and black holes. (NASA Release 86-111)

• Astronaut Sally K. Ride was named Special Assistant to the Administrator for strategic planning, where she would be responsible for reviewing NASA's goals and objectives for near-to-long-term planning. Ride was a mission specialist on two Space Shuttle flights and a member of the Presidential Commission on the Space Shuttle Challenger Accident. (NASA Release 86-114)

• Elizabeth Dole predicted that U.S. rocket companies would begin launching private satellites by 1989 and capture 50 percent of the international market. Her announcement came after President Ronald Reagan's confirmation that NASA would no longer handle the business. Representatives from General Dynamics Corporation and Transpace Carriers, Inc. expressed hope that they would soon announce launch contracts. Some satellite officials, however, were less enthusiastic because NASA had the go-ahead to launch at least some of the 44 commercial satellites, leaving the agency with limited but palpable competition. In a related matter, NASA Administrator Dr. James C. Fletcher said that a large portion of future flights would be devoted to scientific missions and that he expected a commercial launching industry to supplement the Shuttle. (*W Post*, Aug 19/86; *W Times*, Aug 19/86; *NY Times*, Aug 19/86; *CSM*, Aug 26/86)

August 19: The X-Wing Rotor Systems Research Aircraft, which combines the hover characteristics and vertical lift of conventional helicopters and the high cruise speed of fixed-wing aircraft, was formally viewed. Engineers and developers did not foresee the aircraft as a replacement for either fixed or rotary winged aircraft, but one that would serve unique purposes. (NASA Release 86-113)

• Martin Marietta announced plans to enter the private launching business with its Titan class rockets. The Titan was the largest of all U.S. expendable rockets, and spokesman Jack Boyd said that he hoped to market the launch for dual payloads. The corporation had previously submitted a formal request to the Air Force for use of its launch facilities at Cape Canaveral, Florida. (*W Post,* Aug 20/86; *LA Times,* Aug 20/86)

August 20: NASA announced that it would cease implementation of work package realignment for a period of up to 90 days while two teams composed of NASA personnel, Phase B contractors, and representatives from user groups and international partners reviewed Space Station design and work package assignments and functions. The task force, headed by W. Ray Hook, Manager, Space Station Office, would critically examine all aspects of the current Space Station baseline configuration in terms of the amount of extravehicular activity required for assembly and maintenance, launch capacity of the Shuttle fleet, assembly sequence of the baseline configuration, resultant impact to the utilization of the station, potential impact on international partners, and overall technical performance and integrity of the Station. The Executive Technical Committee, headed by NASA Associate Administrator Andrew J. Stofan, would approve the assumptions, engineering, and technical constraints identified by the task force and oversee their activities. (NASA Release 86-116)

• NASA announced that Lieutenant General Forrest S. McCartney, on assignment from the Air Force, would take over as Director of the Kennedy Space Center in Florida, beginning October 1. McCartney succeeded Richard Smith, who retired on July 31. The press opined that the move would give the military greater control of the space program. McCartney's past Air Force assignments include positions as satellite controller, as project officer in the Titan 3 program, in the Directorate of Space at USAF Headquarters, as monitor for satellite communications programs, and director of range engineering at USAF Eastern Test Range. (NASA Release 86-117; *W Post,* Aug 21/86; *NY Times,* Aug 21/86)

August 25: For the fourteenth time since August 1985, the launch of a weather satellite that the Air Force had hoped to put into orbit was again postponed, this time because of a leak of liquid oxygen. Past delays were caused by technical problems with the Atlas-E booster rocket and its payload and scheduling conflicts. The satellite was next intended to be launched from Vandenberg Air Force Base in California on August 30. (*LA Times,* Aug 26/86; *NY Times,* Aug 26/86)

August 27: A NASA Aries sounding rocket was destroyed by a range safety officer at White Sands Missile Range, New Mexico, 50 seconds after liftoff at an altitude of 77,000 feet, when a problem in the vehicle's guidance system was discovered. The rocket carried a 2,300 pound x-ray telescope, designed to study distant stellar objects emitting soft x-rays, jointly developed by

Dr. Gordon Garmire of Pennsylvania State University and Drs. R. Novick and William Hain-Min Ku of Columbia University. (NASA Release 86-119; *W Post,* Aug 28/86)

• The Official Flight Kit and Personal Preference Kits that were recovered during the mission 51-L (Challenger) salvage operation were returned to the institutions, organizations, and families of the crew they were flown for. In addition, NASA planned to present to each state and territory one 51-L crew patch, one U.S. flag, and the respective state or territory flag with the request that these items, which were flown, be displayed appropriately in memorial to the crew. (NASA Release 86-120; *B Sun,* Aug 28/86)

• The American Institute of Aeronautics and Astronautics announced that NASA's budget should be doubled to ensure that the United States retains its lead in space programs. The Institute described the budget as "no growth" and said that it would "insure that the U.S. becomes and remains a second-class power in space." The report called attention to advancements made by the Soviet Union, Western Europe, and Japan, and also recommended attempting to develop a greater student interest in space science. (*B Sun,* Aug 28/86; *W Times,* Aug 28/86; *CSM,* Aug 28/86; *P Inq,* Aug 28/86)

August 28: The U.S. Marine Corps salvage operation for the 51-L Space Shuttle accident, which involved the Freedom Star, the Liberty Star, and the Independence vessels, came to an end after a seven-month effort. It was the largest ocean recovery operation in history and at its peak involved 22 ships and 6,000 personnel, as well as numerous aircraft. The search yielded 45 percent of the Challenger, 50 percent of the external tank and solid rocket boosters, 95 percent of the Spartan-Halley spacecraft, 35 percent of the Tracking and Data Relay Satellite, and 90 percent of the Inertial Upper Stage. (NASA Release 86-121; *C Trib,* Aug 29/86; *B Sun,* Aug 29/86; *W Times,* Aug 29/86; *W Post,* Aug 31/86)

• The Air Force was forced to destroy an unarmed Minuteman 3 ICBM over the Pacific Ocean after an undisclosed problem developed during its 30-minute test flight. Since February 1995, ten successful minuteman test launches had been done when another rocket had to be destroyed. (*NY Times,* Aug 29/85; *P Inq,* Aug 29/86)

August 29: The UNISTICK control system, used in NASA's lunar rover vehicle during the Apollo moon landings, was utilized as a hand-operated device for steering, acceleration, and brake control in vehicles for handicapped drivers. The joystick type system was developed by the Johnson Engineering Corporation, which had made plans for commercial production. (NASA Release 86-123)

August 31: The Peoples Republic of China announced that it was nearing human space flight and was in the process of choosing astronauts for its planned Space Shuttle. It had built and was using an extensive training center and had solved many of the problems space has to offer, such as maintaining life support. No date was given as to when a craft would be built. (*W Post,* Sep 1/86; *W Times,* Sep 1/86)

• NASA said it had canceled 18 Space Shuttle missions over a 5-year period for Spacelab because of the Challenger accident. The agency further stated that it planned to release a list of other scrubbed missions. (*W Post,* Sep 1/86)

September

September 3: NASA Administrator Dr. James C. Fletcher said that the Nation was in need on a fifth orbiter, the funding for which would need to come from private financing. The claim by NASA's advisory council that the United States was approaching second-class status in space, he acknowledged, was "on track". Fletcher made a formal request on September 9, the same day that President Ronald Reagan petitioned Congress for the initial $272 million needed to begin work on a fourth Shuttle. (*USA Today,* Sep 4/86; *NY Times,* Sep 9/86; *B Sun,* Sep 10/86)

• Because of the Challenger accident and the absence of any Shuttle activity, NASA was expected to announce layoffs at its Kennedy Space Center in Florida, where 800 to 1,000 employees would lose their jobs. For the same reason, Martin Marietta said that 700 to 800 of its workers would be laid off, by October 3, from its plant that produced Shuttle fuel tanks. (*W Post,* Sep 4/86; *NY Times,* Sep 5/86)

September 4: Martin Marietta Corporation announced that it had signed an agreement to launch a communications satellite for Federal Express from Cape Canaveral, Florida, in 1989. Because the Titan 3 rocket to be used could deliver payloads up to 32,000 pounds, the corporation was searching for another customer to join the launch. Reagan administration officials were quick to point to the agreement as the birth of the private rocket industry. (*W Post,* Sep 5/86; *P Inq,* Sep 6/86; *LA Times,* Sep 5/86)

September 5: NASA and the American Institute of Aeronautics and Astronautics planned a symposium on Quality and Productivity for December. The event would focus on strategies for revitalizing organizations and provide a forum for discussing issues involved in increasing national quality and productivity. The symposium would bring together over 900 key executives from industry, government, and academia. (NASA anno)

• NASA launched its first successful rocket since the Challenger accident, a Delta rocket from the Kennedy Space Center, Florida. Carrying two payloads for the Strategic Defense Program, one satellite would attempt to track down and intercept the other. McDonnell Douglas Astronautics Company, manufacturer of the rocket, viewed the event as a stepping stone that might lead to the company's entry into commercial launches. NASA had two more Delta rockets in its inventory, one committed to a U.S. Government weather satellite identical to the one lost in May and the other to an Indonesian weather satellite. (NASA Release 86-32; *WSJ*, Sep 8/86, *P Inq*, Sep 6/86; *NY Times*, Sep 6/86; Sep 8/86; *W Post*, Sep 6/86; *C Trib*, Sep 6/86; *LA Times*, Sep 6/86)

• NASA extended a $1.32 billion contract to Lockheed Space Operations Company for processing services at the Kennedy Space Center in Florida. The contracted price, however, did not reflect the schedule changes brought on by the Challenger accident; the changes would be handled through contract amendments. NASA also awarded contracts at $500,000 million each to five different firms for the study of designs for second generation solid rocket boosters. The companies were required only to submit concepts for a newly designed rocket. (*LA Times*, Sep 6/86; *WSJ*, Sep 8/86)

• NASA cleared a site at Wallops Island, Virginia, for use by Space Services, Inc., a new rocket company, to launch its Conestoga rocket. The company planned to put cremated remains of 10,000 people on a satellite that was to orbit the Earth. (*P Inq*, Sep 9/86)

September 8: Boeing Aerospace Company was selected for a one-year contract with two one-year options, valued at $1 million annually, to encourage U.S. firms, with private investment, to join with NASA to explore the application and utilization of new and innovative technologies for the mutual benefit of both parties. Issac T. Gillam, NASA Assistant Administrator, Office of Commercial Programs, noted, "Use of the microgravity environment of spaceflight for research leading to commercial development, production, and application of technologies and materials needed on Earth is a key to the Nation's future in the world economic market." Boeing, familiar with the needs and procedures of both the firm and NASA, would provide the technical and business services necessary to help commercial users prepare proposals to NASA and act as an intermediary. (NASA Release 86-127)

September 9: The first conference on control/structures interaction technology for large, flexible spacecraft, sponsored by NASA and the Department of Defense, was scheduled for November 18–20, 1986. (NASA anno)

• NASA announced the formation of an operations task force for the Space Station, co-chaired by Carl B. Shelly and Dr. Peter T. Lyman. The task force was to explore alternative approaches to operating and managing a deployed Space Station that had to integrate a diverse set of U.S. and International hardware elements to accommodate a wide range of tended and untended user activities and to recommend an effective concept for operating the system. (NASA Release 86-129)

September 10: Astronaut Frederick H. Hauck, former Navy Captain, was appointed Acting Associate Administrator for External Relations. He was commander of STS-51A and pilot for STS-7, as well as a member of the support crew for STS-1 and re-entry capsule communicator on the support crew for STS-2. In mid-October, Kenneth S. Pedersen would become the Deputy Associate Administrator, upon his return from Georgetown University. Since 1979, he had been Director, International Affairs Division and Assistant Associate Administrator for External Affairs. Hauck's duties included policy level management and direction and coordination of the Agency's relationships with public and private organizations. (NASA Release 86-130)

September 11: COSPAS/SARSAT, an international search and rescue system, was scheduled to monitor the around-the-world, non-stop and unrefueled flight of the Rutan Voyager airplane. The SARSAT system uses four satellites, three from the Soviet Union and one from the United States, to locate downed aircraft and ships at sea by picking up their emergency signals. The Rutan Voyager aircraft, designed by aeronautical engineer Burt Rutan and piloted by his brother Dick and Jeana Yeager, carried a 1-1/2-pound prototype beacon that could transmit on a 406 Mhz frequency for location determination and a 121.5 Mhz frequency on which rescue workers could hone in. The signals could also be recorded by a satellite if the aircraft was not in immediate range of a ground station. Since the initiation of the program in September 1982, 606 lives had been saved, by September 1986, in 251 emergencies worldwide. (NASA Release 86-132)

September 15: Fairchild Space Company proposed a lightweight refueling device to be attached to the Space Shuttle's bay that could service satellites with the fuel required for keeping them in orbit. The refueling device would extend the maximum life span for a satellite about 15 years. Fairchild was in a race with Rockwell International and Martin Marietta for the lightest weight and best functioning design; submission of the designs to NASA was to be no later than November 7. (*W Post,* Sep 15/86)

September 16: NASA's Jet Propulsion Laboratory (JPL), in Pasadena, California, studied a program that would send a nuclear-powered spacecraft hurling into deep space traveling 225,000 miles per hour. The mission, said JPL Director Lew Allen, could be launched by 2005. The craft was expected to travel 1,000 astronomical units—1,000 times the distance between the Earth and Sun—sending back data for more than 50 years. The Voyagers 1 and 2 spacecraft were expected to eventually reach deep space, but their slow rates of speed, about 36,000 miles per hour, would put any data gathering activity too far in the future. (LA Times, Sep 16/86)

• The Senate Appropriations Committee approved a plan for a new orbiter that diverted $2.9 billion in defense funds for a reserve to be used by NASA. The proposal asked the Navy to reduce production of the F/A18 fighter and delay construction of a Trident submarine, the Air Force to delay construction of the C17 transport plane, the Army to buy fewer M1 tanks, and defense contractors to shoulder the burden of the cost of tools in its manufacturing. Although the plan called for NASA to oversee construction, the Air Force would launch and operate the new Shuttle. (*W Post*, Sep 11/86; Sep 12/86; *WSJ*, Sep 11/86; *LA Times*, Sep 17/86)

September 17: NASA had another successful launch; a 25-year-old rebuilt Atlas-E rocket, a one-time ICBM, carried a National Oceanic and Atmospheric Administration weather satellite, NOAA-10, into orbit from Vandenberg Air Force Base in California. NOAA Director Larry Heacock said that the polar orbiting satellite would render "half of our satellite program restored to full working capability." (Public Affairs Plan For NOAA-G Launch; *C Trib*, Sep 18/86; *W Post*, Sep 18/86; *W Times*, Sep 18/86; *NY Times*, Sep 18/86)

September 22: Six days after launch and 24 hours after being put into operation, the Search and Rescue Satellite Aided Tracking satellite (SARSAT) picked up its first distress signal from 4 downed Canadians in a remote area of Ontario. Rescue workers were easily able to locate the downed Cessna, and all four Canadians were taken to safety. Canada, France, the United States, and the former Soviet Union all shared in the search and rescue program known as COSPAS/SARSAT aboard the NOAA-10 satellite. (NASA Release 86-144)

September 25: NASA announced to Congress plans for modifying the Space Station that Administrator Dr. James C. Fletcher predicted would be finished by 1994. The new design cut the amount of space walks during construction in half, increased the amount of living and laboratory space, and kept all human aspects concerning the project at the Johnson Space Center, Houston, Texas. The Space Station, according to plans for assembly, would need 17 Space Shuttle flights between 1993 and 1994.

In the meantime, Congress approved legislation, based on a June 4 recommendation, that would require NASA to seek additional manufacturers for solid rocket boosters. The House also voted to authorize spending whatever money would be needed to begin construction of a new orbiter in fiscal year 1987. (P Inq, Sep 26/86; WSJ, Sep 26/86; W Post, Sep 26/86; W Times, Sep 26/86; NY Times, Sep 26/86)

September 26: NASA's Lewis Research Center in Cleveland, Ohio, awarded a $6.6 million, four-year contract to General Electric for definition, modification, and test of propulsion systems for supersonic short takeoff and vertical landing technology development. In support of next century military aircraft, design and modification of a GE F-110 engine would focus on a fan air collection and valving system by taking compressed air from the engine and redirecting it to vertical lift producing devices. (NASA Release 86-136)

September 29: Westinghouse Electric Corporation and Space Industries Company announced their partnership to build a $250 million, permanently orbiting space facility for scientists to perform research. The 14.5-foot-diameter 35-foot-long cylinder would provide a "shirt sleeve" workplace that would benefit a host of industries. (*P Inq,* Sep 30/86)

September 30: NASA released transcriptions of the pre-launch intercom tape of the 51-L, Space Shuttle Challenger, for public consumption. The tape was about an hour long, beginning when the crew first entered the craft to two minutes and five seconds before lift-off. It contained nothing that was relevant to the accident and was not provided to the Presidential Commission in support of the investigation. Because the intercom channel provides a private mechanism for exchanging spontaneous comments and recording information of a personal nature, NASA had not released these transcripts previously. Because of media inquiries, however, the pre-launch tape was transcribed. (NASA Release 86-135; 86-137; 51-L Transcript Folder; *NY Times,* Oct 1/86)

October

October 1: The European Space Agency announced that it planned to fly four missions relating to the Space Station project. The maiden flight for the Spaceplane Hermes, said Director General Dr. Reimer Lust, would be launched between 1996 and 1998, by a new Ariane 5 rocket. (*NY Times,* Oct 6/86)

October 2: NASA announced that Aaron Cohen would be appointed as Director of the Johnson Space Center (JSC), Houston, Texas, and that Jesse W. Moore, in response to his own request, would be reassigned from that position to be Special Assistant to the General Manager, NASA Headquarters. Since 1962 Cohen had served NASA first in the Apollo Program, then in the Space Shuttle Orbiter project as Director of the Engineering and Development Directorate, and then as overall Director of Research and Engineering at the JSC. Moore came to NASA in 1978 as Deputy Director of the Solar Terrestrial Division, Office of Space Science. From there he was appointed Director of the Space Flight Division; Director, Earth and Planetary Exploration Division; and Deputy Associate Administrator and acting Associate Administrator for Space Flight. (NASA Release 86-138; *NY Times,* Oct 3/86)

• A decision was reached by NASA to test the new booster rocket design for the Shuttle fleet by firing the rockets horizontally, and not vertically as the Rogers Commission had suggested. A NASA spokesman called this method of testing "the most demanding" and the one that "best satisfies the Commission's intent." (*NY Times,* Oct 3/86; *W Post,* Oct 3/86; *P Inq,* Oct 3/86)

• The Jet Propulsion Laboratory Successfully tested a new space radio-astronomy technique, Very Long Baseline Interferometry (VLBI), that combined data from radio telescopes on the ground with data from an antenna on NASA's Tracking and Data Relay Satellite System. The result was a better resolution of three quasars (among the most distant objects known and designated 1730-130, 1741-038, and 1510-089) than what ground-based radio studies at the same wavelength could provide. It was the first time an orbiting satellite had been used as a radio telescope. (NASA Release 86-140; *W Post,* Oct 3/86; *LA Times,* Oct 3/86)

October 3: NASA opted to use the deactivated Minuteman facilities at Cape Canaveral Air Force Station, Florida, for long-term storage of debris recovered from the Space Shuttle Challenger accident. Large concrete covers were to be placed over the tubes for a weather tight seal, concluding NASA's primary activities related to the analysis and disposition of recovered hardware. (NASA Release 86-141)

• NASA announced a seven-year schedule of tasks for the Shuttle fleet, and that Shuttles would resume flying in early 1988. The schedule reflected President Ronald Reagan's directive to phase out commercial and foreign launches, inasmuch as 24 of the 44 contracts had been voided. This left only 12 percent of the Shuttle missions devoted to commercial and foreign launches. An increase to 41 percent of the payloads, as opposed to 31 percent, would be given to Department of Defense missions. The remaining 47 percent of the schedule was reserved for the space agency's own missions, 33 percent for scientific studies and other needs, and 14 percent for the Space Station. The first year, said NASA, would see only five flights; that number would, it was hoped, increase to 16 flights per year by the 1990s. (*WSJ*, Oct 6/86; *W Post*, Oct 4/86; *P Inq*, Oct 4/86; *B Sun*, Oct 4/86; *NY Times*, Oct 4/86)

October 7: A House Science and Technology Committee issued a report stating that Congress and the White House shared the blame for the launch schedule pressure that led to the January 28 Challenger accident. The report agreed with many of the findings laid out by the Rogers Commission, but differed by stating that the accident was not a result of poor communication through NASA management, but "poor technical decision-making [by them] over several years." The report concluded that NASA's management lacked the expertise needed to operate the Space Shuttle and warned the Agency not to succumb to similar pressure concerning future flights.

The National Research Council, in an independent study requested by Congress, urged NASA on October 9 to limit its Shuttle flights in the 1990s to 11 or 13, not the 16 annual flights proposed by NASA Administrator Dr. James C. Fletcher. (Investigation of the Challenger Accident: Report of the Committee on Science and Technology, House of Representatives, Oct 29/68; W Post, Oct 8/86; Oct 10/86; W Times, Oct 8/86; NY Times, Oct 8/86)

October 9: NASA and the Federal Aviation Administration signed a memorandum of agreement for a five-year, $24 million research project to study wind shear detection and avoidance that would cover technology assessment, present position sensor integration, hazard characterization, pilot factors in wind shear, and effects of heavy rain. For pilot training, an environment was created that could mirror wind shear in a flight situation, thus providing a safe simulation for training.

Microburst is the most dangerous type of wind shear for study. It is a small, intense down-draft, often accompanied by heavy rain, which, upon striking the ground, spreads out into a circular vortex radiating an all directions and leaving little time for reaction by a pilot. Between 1965 and 1985, there were at least 26 accidents and 3 incidents involving 626 fatalities and 235 injuries where wind shear was a direct cause or contributing factor. (NASA Release 86-143)

October 12: NASA scientists were searching for ways to clean up the orbiting debris left from nuclear powered satellites and wreckage from anti-satellite weapons testing. Considering the high speed of orbiting craft, scientists reported that even a fleck of paint could be dangerous, the probable culprit that caused a cracked windshield in the seventh U.S. Shuttle flight. Because slightly larger objects could result in catastrophic damage, the planned Space Station sacrificed some valuable weight to armor crew quarters. (*W Times,* Oct 13/86)

October 13: A panel of scientists and engineers, established at the behest of the Rogers Commission, endorsed NASA's redesign of the Shuttle's solid rocket, but urged the Agency to increase its test schedule and to consider alternative designs. NASA's efforts had been limited by confining any redesign to plans that incorporated the rocket segments already ordered before the Challenger accident. Because the chosen design could prove inadequate and because NASA would begin investigating second generation booster rockets, all alternative design changes, the panel concluded, were in need of study. (*P Inq,* Oct 14/86; *W Post,* Oct 14/86; *NY Times,* Oct 14/86)

October 14: NASA announced that it would allow private contracts for McDonnell Douglas, builders of the Delta rocket, and General Dynamics Corporation, builders of the Atlas Centaur, to market their products. The two companies were also granted the use of NASA launch facilities, with the agreement to reimburse for any costs incurred. Talks with Transpace, a competitor who had hoped to market the Delta rocket, were ended. (*W Post,* Oct 15/86; *W Times,* Oct 15/86; *WSJ,* Oct 15/86)

October 15: A balloon launched from the National Scientific Balloon Facility carried a telescope camera 22 miles up to study gamma rays from stars, black holes, and other sources. It was considered the highest resolution telescope for that purpose. (*W Times,* Oct 16/86)

October 16: NASA planned to proceed with construction of a second horizontal test stand for redesign and recertification of the Space Shuttle Solid Rocket Motor (SRM) at the Morton Thiokol, Inc., Wasatch facility in Utah. The newer test stand would more closely simulate the stresses on the SRM during an actual launch and ascent. (NASA Release 86-146; *W Times,* Oct 17/86)

• Communications Satellite Corporation announced that it had found a way to double the life of an orbiting satellite by reducing fuel consumption. Satellites in a geosynchronous orbit appear stationary over the Earth, allowing ground stations to communicate. Because of gravitational forces, most coming from the moon, satellites tend to drift north or south to a degree where communication can be lost. To compensate for the drift, rockets are fired to bring a satellite back to its original orbit. Most of a satellite's fuel is spent firing its rockets. The new approach, elements of which are used by the military, allows the satellite to drift and be tilted by its rockets so that the antenna's "footprint" can be located. Ground stations receiving messages needed to undergo some modifications as well, however. (*W Post,* Oct 17/86; *WSJ,* Oct 17/86; *NY Times,* Oct 17/86)

October 17: Using data collected between June 1978 and summer 1986, from NASA's costal zone color scanner on board the Nimbus-7 satellite, scientists from many international institutions completed a computer-generated color image that shows the distribution of microscopic plant life, phytoplankton, in surface waters over the North Atlantic Ocean. This image was generated by measuring sunlight reflected from the sea surface at five wavelengths to determine ocean color. The most striking feature of the study for understanding man's role in carbon dioxide and other important cycles is the "spring bloom," high phytoplankton concentration extending across the Atlantic and into the North Sea. (NASA Release 86-147)

October 19: NASA released studies done by Donald Heath, a scientist at the Goddard Space Flight Center, Greenbelt, Maryland, which reveal that the Earth's ozone layer is thinning at a much higher rate than previously thought, about three percent between 1978 and 1984. This data was gleaned from a satellite orbiting 570 miles above the Earth. (*C Trib,* Oct 20/86; *B Sun,* Oct 20/86)

October 20: Dr. Dale D. Myers became Deputy Administrator for NASA; he succeeded Dr. William R. Graham who left NASA to become Director of the White House Office of Science and Technology Policy. Prior to this appointment, Dr. Myers had served as an at-large member to the NASA Advisory Council and before that as president and chief operating officer for Jacobs Engineering Group. Other achievements include positions as Undersecretary for the U.S. Department of Energy, vice-president of Rockwell International, president of North American Aircraft, Associate Administrator for Manned Space Flight, NASA, and vice-president/program manager for both the Apollo Command/Service Module Program and the Space Shuttle Program at Rockwell International. (NASA Release 86-149)

October 21: Several senior staff changes were made at the Johnson Space Center, Houston, Texas, to include: Robert C. Goetz, from Deputy Center Director to Technical Assistant to the Center Director; Clifford E. Charlesworth, from Director of Space Operations to Special Assistant to the Director; Henry O. Pohl, from Division Chief to Director of Engineering; William R. Kelly, from Director of Center Support to Director, Administration; Dr. Wayne R. Young was named Deputy Director for Administration; Paul J. Weitz, from Deputy Chief of the Astronaut Office to Technical Assistant to the Center Director. (NASA Release 86-151)

• Thomas L. Moser was named Director of the Space Station Program Office in Washington, D.C., a position created to conduct a long-range assessment of overall NASA capabilities and requirements. Moser began his career with NASA in 1963 as a mechanical systems design and analysis engineer, and served further as a structural subsystem manager for the Apollo command module; project manager for the Shuttle Structures and Mechanics Division; head of structural design and manager for the orbiter structure and thermal protection system; technical assistant to the director and deputy manager for the Orbiter Project Office; Director of Engineering at the Johnson Space Center, Houston, Texas; and Deputy Associate Administrator for Space Flight at NASA headquarters. (NASA Release 86-152)

October 22: The Johnson Space Center (JSC) in Houston, Texas, announced plans to build, and open by 1989, a $40 million visitor's center near its complex. The center would include a large geodesic dome, two high-tech movie theaters, a simulated lunar landscape, a Mission Control Center simulation, and an exhibit hall featuring past achievements. (*LA Times,* Oct 23/86)

October 23: Joseph F. Sutter was appointed chairman of the NASA Aerospace Safety Advisory Panel, an advisory group to the NASA Administrator. Sutter, in 1986, had a 41-year career with the Boeing Company as an aerodynamicist, received numerous awards for development of jet-powered airliners, and was nominated to receive the Wright Memorial Trophy, to be presented in Washington, D.C. He had also been appointed to the Presidentiaal Commission on the Space Shuttle Challenger Accident. (NASA Release 86-154)

October 27: NASA engineers settled on a design for the Shuttle's new booster rockets, and beginning in early 1987, four full-scale tests were planned. John Thomas, manager of the booster design project at Marshall Space Flight Center, Huntsville, Alabama, predicted that the tests would run smoothly and that Shuttle flight would resume in February 1988. (*C Trib,* Oct 28/86)

October 28: In conjunction with its 1986 Small Business Innovation Research Program, NASA announced the selection of 172 research proposals from 144 small firms for Phase I Award Negotiations. Selected from 1628 proposals, about one-half of the six-month, fixed-price contract efforts would be eligible for Phase II follow-on contracts of up to two years. This program was designed to stimulate technological innovation in the private sector and strengthen small business participation. (NASA Release 86-155)

October 30: A study by NASA scientist Linwood B. Callis concluded that the recently discovered thinning ozone layer might have been a result of increased sunspot activity. Other scientists, however, were hesitant to accept the theory and remained steadfast in their belief that manmade chemicals were responsible for the depletion. (*LA Times,* Oct 30/86; *C Trib,* Oct 31/86)

October 31: Parallel contracts for the Advanced Stirling Conversion System were awarded by NASA's Lewis Research Center, Cleveland, Ohio, to Mechanical Technical, Inc. and Stirling Technology Company, for $253,385 and $246,576, respectively. Research was to be funded by the Department of Energy, as ground application took precedence over any potential space uses. The system would include a free-piston Stirling engine, a liquid-metal heat pipe receiver, and a means to provide power to a utility grid. The external combustion engine has only two moving parts and works by heating air, or another gas, in a cylinder, with a power source at one end, such as focused solar energy, which expands the gas and moves the piston. Electric power to the utility grid would be transferred from the engine directly, via a linear alternator, or indirectly with hydraulic output to a ground-based fluid pump coupled to a generator. (NASA Release 86-156)

• NASA announced plans for extending operations to five of its ground tracking stations located at Ascension Island in the southeast Atlantic, Santiago (Chili), Guam, Hawaii, and western Australia. Closing dates for these stations was moved up and was conditional on the February 1988 launch by the Space Shuttle of a second Tracking and Data Relay Satellite (TDRS); a TDRS was destroyed during the Challenger accident. Two TDRS's could provide 85 percent coverage of an orbiting spacecraft, compared with the 50 percent coverage provided by ground stations. (NASA Release 86-157)

November

November 3: Several engineers at NASA recently questioned the proposal to add 100 bolts to a critical joint between the exhaust nozzle and the solid fuel segments in the Shuttle booster rockets. The purpose of the fasteners was to prevent further leakage of hot gases. Many engineers, however, saw these holes as another way for the gasses to escape because they were to be placed between inner and outer o-rings. (*B Sun,* Nov 3/86; *W Times,* Nov 3/86)

November 4: NASA officials announced that James Kingsbury, engineering chief at Marshall Space Flight Center, Huntsville, Alabama, would retire at the end of November. He was one of the high ranking officials who had worked on Challenger's fatal rocket boosters. (*USA Today,* Nov 5/86)

November 5: In a part of NASA's strategy to return the Space Shuttle safely to flight, Dale D. Myers, Deputy Administrator, announced the new management and operations structure for the National Space Transportation System (NSTS or Space Shuttle program). The changes were aimed at establishing clear lines of communication in the information transfer and decision making processes. The management change, which also moved control of the operation to Washington, D.C., was directed by Shuttle Director Richard H. Truly and followed recommendations made by the House of Representatives Committee on Science and Technology and the Presidential Commission on the Space Shuttle Challenger Accident.

As part of the restructuring, Arnold D. Aldrich, who joined NASA in 1959, was named Director, NSTS, and Richard H. Kohrs and Robert L. Crippen were named his deputies. Truly stated that "the formation of this program management organization is one of the most positive steps NASA has taken in the last several months to strengthen the Shuttle program and return to flight status." (NASA Release 86-159; *W Post,* Nov 6/86; *NY Times,* Nov 6/86)

• The United States and the European Space Agency were pursuing an agreement aimed at preventing Japan from dominating small satellite launches in the same way it had taken over the small car industry. The United States hoped to reach assent with the European Space Agency concerning employment schedules, insurance standards and rates, Government subsidies for research and development, and the leasing costs of Government-owned launch facilities to private firms. (*W Times,* Nov 6/86)

November 7: NASA's Ames Research Center dedicated the $2.3 million, 21,336-square-foot Fluid Mechanics Laboratory, a new research facility designed to stimulate basic knowledge of aerodynamic flows. The Laboratory could support multiple, simultaneous small-scale wind tunnel research while having full access to supercomputer resources. The laboratory would concentrate on vortex flows, a major factor in maneuverability and reduced drag by aerodynamic means, for applications ranging from developing fuel-efficient aircraft to low-speed research on the National Aero-Space Plane. (NASA Release 86-158)

November 8: United Technologies Corporation proposed implementing the launch propulsion and guidance system used in NASA's Space Shuttle for an untended space vehicle. The guidance system, they averred, would allow untended rockets to deliver much heavier payloads at a cheaper cost than the Shuttle. Operation, they claimed, would begin about three years after NASA approved the plan. (*AP,* Nov 8/86)

November 10: Negotiators concluded an agreement between the United States and the Soviet Union for cooperation in scientific exploration, primarily focusing on Mars research, said officials from NASA. The agreement, said other sources, made no commitment to a specific joint mission and could be signed at the upcoming summit meeting. (*LA Times,* Nov 11/86; *CSM,* Nov 11/86)

• NASA Administrator Dr. James C. Fletcher would likely seek to continue the $10 billion budget afforded to fiscal year 1987, a 40 percent increase from fiscal year 1986. Fletcher, according to NASA officials, planned to meet with President Ronald Reagan and argue that the Challenger accident was a result of years of underfunding. The increased funds could support the Civil Space Technology Initiative for promoting space research and technology; the Global Geoscience System, a solar terrestrial satellite new start; High Alpha, a military aeronautics research program; and the X-ray Astrophysics Facility, advanced technology development. The increased NASA budget resulted from a $2.4 billion decrease in the Department of Defense budget. (*AvWk,* Nov 10/86)

• Technology developed by NASA for delivering a culture medium into the Martian soil was modified and implanted into a diabetic patient to serve as an insulin pump. Although other insulin pump implants have been used in the past, the space-age pump not only had a longer life, but also was programmable after being implanted. (*C Trib,* Nov 20/89)

November 11: A week-long exercise to upgrade the Search and Rescue Satellite System got underway at the National Oceanic and Atmospheric Administration, expanding coverage to the Southern Hemisphere. (LA Times, Nov 12/86)

November 13: Martin Marietta, in competition for a contract to build the Space Station, unveiled a full-scale model depicting its plans at Marshall Space Flight Center, Huntsville, Alabama. The intended crew module has eight sleeping quarters, each with a video recorder and a computer giving crew members direct access to home and the Space Station's main computer. (*C Trib*, Nov 14/86)

• A $13 million Air Force satellite that had served as a museum piece at the Smithsonian Institution was renamed Polar BEAR and lifted into polar orbit aboard a Scout rocket, NASA's smallest booster. The satellite's mission was to photograph the Northern Lights and sample electrical particles and the magnetic field in order to design better communications. (*NY Times*, Nov 15/86)

• The third stage of a French Ariane rocket that had been launched nine months earlier blew up on this date. U.S. officials asked Arianespace to look into the matter in order to prevent a recurrence because orbiting debris posed an increased danger to other satellites. (*W Times*, Dec 2/86; *W Post*, Dec 2/86)

November 14: President Ronald Reagan vetoed a NASA authorization bill on the grounds that the bill's attempt to create a National Space Council (NSC) interfered with Presidential decision-making power. Establishment of the NSC, he said, would duplicate other interagencies while creating "unacceptable interference with my discretion and flexibility." The move did not affect NASA's budget. (Memorandum Withholding Approval of H.R. 5495, Nov 14/86; *P Inq*, Nov 15/86; *NY Times*, Nov 15/86)

November 18: For the first time since the Challenger accident, astronauts boarded a Space Shuttle; they took part in a mock countdown at Cape Canaveral, Florida, on the Atlantis. Computers halted the drill 25 seconds short of simulated engine firing, but officials deemed the test successful. (*W Post*, Nov 19/86; *NY Times*, Nov 19/86)

November 21: Leaders from the 11-member-nation European Space Agency (ESA) met with NASA officials to discuss their $2 billion contribution to the Space Station project and to seek a greater management role and benefits from it. One European official noted, "NASA must realize that ESA is a mature space agency that can manage manned space flight alone if necessary." Although the statement appeared to be a threat to withdraw, Richard Barnes, NASA Director of International Affairs, saw no intent on the part of Europe to pull out. Leaders from both sides, however, agreed to a joint statement saying, "While recognizing that differences currently exist, Professor Luest and Dr. Fletcher jointly concluded that these differences are not irreconcilable." (*NY Times*, Nov 22/86; *AvWk*, Nov 24/86)

November 24: The results of a study conducted by Geologists Dr. Matthew Golombek and Dr. Laurie Brown suggested that the northwest Mojave Desert had shifted about 25 degrees clockwise because of periodic, violent earthquakes some 16 to 20 million years ago. Golombek reported that "results of magnetic studies of volcanic rocks taken from 19 sites suggested that the movement was caused by shear (the Earth's tectonic plates sliding past or over one another) from the Pacific Plate sliding along the fault past the North American Plate." Magnetic minerals had once lined up parallel with the Earth's magnetic field to the North, but the study found minerals now pointing 25 degrees to the east, and it was determined that tectonic movement had rotated the rocks. (NASA Release 86-160; *LA Times,* Nov 30/86)

November 26: NASA selected the Inertial Upper Stage (IUS), a launch vehicle that fits in the cargo bay of the Space Shuttle, to carry three probes for planetary missions. Despite earlier plans for alternative launch vehicles following cancellation of the Shuttle/Centaur Upper Stage and the Challenger accident, NASA felt an urgent need to reestablish its planetary program. The probes scheduled for launch in 1989 and 1990 included the Galileo, a joint mission with Germany that would circle Jupiter and measure electromagnetic fields and plasma particles; the Magellan, which would orbit and map Venus using radar; and the Ulysses, a joint mission with the European Space Agency designed to orbit and study the poles of the sun. These three were the first missions for orbit around bodies other than Earth that would employ the IUS.

The IUS boosters were manufactured by Boeing to launch three scientific missions. They were to have launched the General Dynamics hydrogen-fueled Centaur rockets, but the highly volatile boosters, thought possibly unsafe for human space flight, were grounded after the Challenger accident. Because of its less powerful rocket, the Galileo needed to slingshot Venus in order to reach Jupiter, as opposed to flying directly to the planet. The added time for the mission was expected to run $50 million. (NASA Release 86-161; *NY Times,* Nov 27/86)

November 28: NASA's Marshall Space Flight Center in Huntsville, Alabama, awarded an $8.9 million five-year contract to Calspan Corporation for operation of the Marshall Problem Assessment Center and the Safety Issue Assessment Center. Under the contract, Calspan would receive problem reports and safety issues from the contractors of Marshall's numerous projects, check the accuracy and completeness of the reports, track the closeout of proposed corrective actions, perform trend analysis and special study into causes of problems and safety issues, and appraise Marshall management on a daily basis. (NASA Release 86-163)

• Marshall Space Flight Center, Huntsville, Alabama, announced four key appointments in the Science and Engineering Directorate, in an effort to meet the goal of getting the Space Shuttle back to full operation. Dr. Judson A. Lovingood was named Associate Director for Propulsion Systems, with responsibility for propulsion projects for the Space Shuttle Main Engine, Solid Rocket Booster, External Tank, Orbital Maneuvering Vehicle and upper stages. E. Ray Tanner was named Associate Director for Space Systems, with responsibility for assuring engineering adequacy of the Space Station, Hubble Space Telescope, Advanced X-Ray Astrophysics Facility, and Marshall Center-assigned payloads and Spacelab payload integration. John P. McCartney was named Director, Propulsion Laboratory to oversee research and development, engineering, and technical direction of propulsion systems design, and analysis related to launch and space vehicles. Dr. George F. McDonough was named Director, Structures and Dynamics Laboratory with responsibility for research and development in structural design and analysis of launch and space vehicles, analysis dynamics behavior, specification of dynamics-related design criteria, and analysis of atmospheric and environmental processes. (NASA Release 86-164)

• A Tracking and Data Relay Satellite lost transmitting capability on an S-band single-access antenna, SA-1, which provided voice and data links to low Earth-orbiting satellites, including the Shuttle. Services were switched to the remaining antenna, SA-2, and little or no data loss to customers occurred. (NASA Release 86-175)

November 30: Astronaut Joe H. Engle retired from the United States Air Force as a colonel and resigned from NASA. After coming to NASA in 1966, he commanded Columbia and Discovery flights and was named Deputy Associate Administrator, in March 1982, for the Office of Space Flight. (NASA Release 86-166; *W Post,* Dec 3/86)

December

December 2: The Soviet Union recently had their version of the Space Shuttle on a launch pad for tests. Photos of the incident were taken by a U.S. satellite, and Government officials predicted that the country would begin human space flight in 1988. (*B Sun,* Dec 2/86; *W Post,* Dec 2/86)

Western European officials announced that the grounded Ariane rocket would not be ready to fly in February as was earlier hoped, adding to the strains fed by the West's inability to launch. Modifications were recently approved for the rocket's faulty third stage; no new flight date was given. (*NY Times,* Dec 3/86)

December 3: McDonnell Douglas Astronautics was selected for negotiations leading to the award of a three-year contract, with another three-year option, and a six-year price tag of $327 million, to perform Payload Ground Operations for Kennedy Space Center activities, both in Florida and elsewhere inside and outside the United States. The work was being done by five different contractors, and NASA hoped to obtain a single, long-term contract. Responsibilities included payload/cargo processing and integration, Spacelab operations and integration, support to experimental integration activities, payload/cargo de-integration, NASA/Vandenberg payload operations, payload related facilities, systems and ground equipment operations, maintenance and sustaining engineering, customer accommodation and launch-site support functions, and payload related support operations and services. (NASA Release 86-167)

• NASA participated in a study of controlled California forest fires in the San Gabriel mountains. The data focused on the global effects of fire on atmospheric quality, air and water pollution, erosion, soil depletion, and species extinction, as well as how a fire could be handled more effectively. Scientists hoped to learn more about the effects of fire on different biogenic gases, such as nitrogen oxide and methane hydrocarbons, and how changes in these gases affect the atmosphere, the climate, and the biosphere. For its aerial study, NASA flew a U-2 aircraft at 60,000 feet. A similar study on the effects of a nuclear winter was scheduled. (NASA Release 86-134; *W Post,* Dec 4/86)

December 4: Remote sensing via satellite, combined with a computer technology that could provide a simulation of the properties and processes in the soil, related Dr. Elissa Levine, would allow short-term prediction of changes that occur in soils over a period of time and the ability of the soil to support different types of plant life. Advancing her theory at an annual meeting in New Orleans, Levine noted, "We can obtain critical environmental data which will help us understand how changes brought about by both nature and humankind will affect the total ecosystem over time." (NASA Release 86-170)

• The sixth FLTSATCOM communications satellite, F-7, was launched by an Atlas-Centaur rocket fired from Cape Canaveral, Florida. The Department of Defense added this satellite to their communications network among naval aircraft, ships, submarines, ground stations, Strategic Air Command elements, and Presidential command networks. With this addition, the network could provide 23 high frequency communication channels and one super high frequency channel with an experimental Extra High Frequency package. The rocket had been grounded since May because of previous NASA rocket failures, beginning in January with the Challenger explosion, and had experienced eight delays before becoming NASA's third successful major flight of the year. (NASA Release 86-165; *W Post,* Dec 5/86)

• A joint venture of Hughes Aircraft and RCA, the Earth Observation Satellite Company said that the Reagan administration was withholding funds for the Landsat program it managed. The $65 million not released was earmarked for new satellites, receiving stations, and support operations. The Landsat satellite was launched in 1972 as a data gathering operation for the government and public, but its monopoly for providing detailed services to companies and private individuals was broken by the French satellite, SPOT. Elimination of the program, warned a representative from the company, "would destroy the first U.S. attempt to commercialize space." (*W Post,* Dec 5/86)

December 7: NASA Administrator Dr. James C. Fletcher, responding to negotiations with the White House budget office earlier in the week, warned that the planned $8 billion Space Station and a new replacement Shuttle were in jeopardy of not meeting scheduled completions. He stated that the 35 percent increased budget for fiscal year 1987 was earmarked only for costs resulting from the Challenger crash and could not be benchmarked for future budgets. The strict budget, he added, could afford no new projects.

In a separate statement, Fletcher agreed to remove himself from any future Shuttle booster rocket contracts. Senators called for an investigation of the contract awarded to Morton Thiokol in 1973, a decision in which Fletcher had considerable influence. (*W Times,* Dec 8/86; *NY Times,* Dec 8/86)

December 8: NASA looked ahead to the launching of four new generation space observatories that promised a new era for astronomy. The Gamma Ray Observatory, providing a greater wavelength range, could determine if arriving gamma radiation originated in quasars and pulsars or from other sources. The Advanced X-Ray Astrophysics Facility would study highly energetic environments found in nearly every object in the universe, including stars, planets, neutron stars, black holes, quasars, and cores of active galaxies. The Space Infrared Telescope Facility could span the infrared part of the spectrum with a 1000-fold increase in sensitivity, enabling it to search for planets around stars. The Hubble Space Telescope, penetrating the universe in visible and ultraviolet light, was predicted to expand the observable universe by hundreds of times, see objects with ten times the clarity of ground observatories, and detect objects 1,000 times dimmer than those observed by previous spacecraft. (NASA Release 86-168)

• NASA's Jet Propulsion Laboratory in Pasadena, California, developed an airborne sensing device, an infrared radiometer, for measuring ocean surface temperatures and charting temperature and wind maps. Tested over southern California's ocean aboard the Goodyear blimp Columbia, the new instrument was unique in that it could distinguish water temperature from the temperature of the air immediately above it. Researchers expected that the device could greatly enhance satellite capability to monitor ocean weather. (NASA Release 86-169)

• Because the ocean surface mirrors the ocean floor as gravity pulls water down into depressions and forces it up around the mountains (looking north from Puerto Rico, for example, the ocean surface drops nearly 60 feet), scientists at the Goddard Space Flight Center, Greenbelt, Maryland, employed satellites to map the ocean's floor. Data taken from altimeters on the Geodynamic Experimental Ocean Satellite and the Sea Satellite were used to generate a computer image of the ocean surface, reflecting the Earth's structure underneath the water. The data, said Dr. James Marsh, also increased knowledge of circulation and current systems. (NASA Release 86-172)

December 9: Forty-nine research proposals from 45 small, high technology U.S. firms, totaling about $23 million, were selected for immediate negotiation of Phase II contract awards. These proposals were chosen from 100 submitted to the Small Business Innovation Research program, a program designed to strengthen the role of small business participation in Federal research and development and contribute to the growth and strength of the private sector. (NASA Release 86-174)

• NASA signed an agreement with the Minnesota Mining and Manufacturing Company (3M) allowing the company to fly 62 materials processing experiments aboard future Shuttle flights. The microgravity experiments in organic and polymer science were to be conducted on a space-available basis, and NASA would have use of 3M's experiment equipment for scientific investigations. (*UPI,* Dec 9/86)

December 12: NASA scientists halted design of additional shielding for two nuclear powered satellites. The shield, they ascertained, would help only during the first two minutes of launch and would actually become a detriment after that. The Interagency Nuclear Safety Review Panel would not give launch approval for the scientific missions until 1999, but NASA officials saw a good chance for approval of launching the radioisotope thermoelectric generator powered satellites from the Space Shuttle. (*AvWk*, Dec 15/86)

December 16: NASA's Ames-Dryden Flight Research Facility in Edwards, California, modified an F-15 jet engine to increase its thrust by 10 percent and fuel savings by 5 to 7 percent. The modification grew out of NASA's Highly Integrated Digital Electronic Control program (HIDEC), which lessened unneeded engine stall margins (the amount that engine operating pressure must be reduced to avoid stall). The typical 25 percent stall margin allows for a contingency of any of the worst flight conditions and robs 15 percent of an engine's usable power. The HIDEC created engine and flight control systems that communicate with each other; thus the engine adjusts itself according to actual flight, not assumed constants. (NASA Release 86-176)

• General Samuel Phillips presented the Report of the NASA Management Study Group, for which he was Project Director, to Administrator Dr. James C. Fletcher. The report conducted an assessment of NASA practices in long-range and strategic planning, managing NASA programs, human resources and procurement, maintaining budget, financial and program control, and dealing with external affairs. The report went on to assess the effectiveness of NASA's organization and to make recommendations to improve agency practices and organization. (Report of the NASA Management Study Group, Dec 16/86)

December 18: Senator Albert Gore made public his order for a congressional investigation into whether NASA Administrator Dr. James C. Fletcher had violated conflict-of-interest rules when he awarded Morton Thiokol the Shuttle booster contract in 1973. Dr. Fletcher denied the charges, but said he would seriously consider removing himself from future booster rocket contract decisions. (*NY Times*, Dec 19/86)

December 19: The Pentagon, which had previously voiced objection to the Space Station, asked for a postponement in negotiations between NASA and foreign participants because the Department of Defense (DoD) hoped to use the orbiting laboratory for some of its Strategic Defense Initiative experiments. The DoD envisioned only a very limited need for the facility, but other countries would no doubt vehemently object because the project was intended only for peaceful purposes. (*WSJ*, Dec 22/86; *NY Times*, Dec 20/86)

December 23: Analysis conducted by the Office of Space Station in the areas of Space Station management, use of expendable launch vehicles, and cost impacts resulting from design changes was accepted by NASA. Changes included replacing the nodes and tunnels that connected pressurized modules with larger resource nodes, revising the assembly sequence to provide early scientific return and reduce extravehicular activity, achieving an initial power level of 37.5 kilowatts, achieving permanent human capability, and placing fixed servicing capabilities closer to the modules. Realignment of certain work package responsibilities was also recommended; the total cost increase would be around $49 million. (NASA Release 86-180; 86-181)

December 27: NASA ruled out a recommendation to supplement its Shuttle flights with untended launch vehicles for building the Space Station. The combination would accelerate assembly by up to nine months, said Space Station manager John Dunning, but the time savings would be offset by a 10 to 40 percent increase in potentially dangerous space walks. Using expendable launch vehicles, he went on, would also require an orbital maneuvering rocket to keep the parts in place. (*NY Times,* Dec 28/86)

December 31: Astronaut Paul J. Weitz was named deputy director of the Johnson Space Center (JCS), Houston, Texas. Past achievements include participation in the first astronaut visit to the Skylab, serving as commander of the sixth Space Shuttle mission, and serving as technical assistant to the JSC Director. (NASA Release 86-183)

• NASA extended a one-year, $129.3 million contract to EG&G Florida, Inc. for base operation services, bringing the total to $565.3 million. The extension called for EG&G to provide institutional and technical support services at the Kennedy Space Center, in Florida. (NASA Release 86-184)

• The Marshall Space Flight Center, Huntsville, Alabama, selected Computer Sciences Corporation for negotiations leading to a 10-year, $99 million contract to operate its computer facility. The company would provide the personnel, equipment, and supplies necessary to carry out duties at the Slidell Computer Complex. (NASA Release 86-185)

During 1986: The Space Shuttle Challenger accident and the subsequent investigation and recovery activities headed the list of prominent events for NASA. Some major accomplishments, however, are worthy of note: Voyager 2 encountered the planet Uranus in January, continued with its journey and was expected to reach Neptune by 1989. A new baseline configuration was drafted for the Space Station—a major milestone for development. NASA and the Department of Defense initiated a joint Aero-Space Plane research program concerning vehicles capable of horizontal takeoff and landing, single-stage operations to orbital speeds, and sustained hypersonic cruise within the atmosphere using air breathing propulsion. NASA signed an agreement with the 3M Company for conducting 62 materials processing experiments aboard the Space Shuttle, and Dr. James C. Fletcher, for the second time, became Administrator of NASA.

Two books concerning NASA endeavors were published in 1986. Joseph P. Allen's *Entering Space* deals with an astronaut's journey into the last frontier, and William E. Burrows' *Deep Black* looks into space espionage and national security. (NASA Release 86-177)

January

January 5: Jet Propulsion Laboratory (JPL) Director Lew Allen announced that Richard P. Laeser had been appointed to manage JPL's new support office in Washington, D.C., for NASA's Space Station Program. Laesar, who until his new appointment had been the manager of the Voyager mission to outer planets, would also serve as Director of the Office of Requisitions and Assessments for the Space Station.

Norman R. Haynes was appointed the new manager of the Voyager mission. He formerly was manager of JPL's Systems Division. (*Star-News,* Jan 5/87)

January 6: NASA announced that the Cosmic Background Explorer (COBE) satellite would be launched on a Delta expendable launch vehicle in early 1989 rather than on the Space Shuttle in July 1988, as originally planned. The change in schedule was caused by the backlog of science payloads awaiting launch on the Space Shuttle as a result of the Challenger accident. Designed to study the "Big Bang", COBE would be launched into a 560-statute-mile, sun-synchronous orbit from Vandenburg Air Force Base in California. Because of the switch from the Shuttle to Delta, COBE would be reduced in weight from 10,500 pounds to 5,000 pounds and in size from 15 feet to 8 feet in diameter. Scaling down the spacecraft would also require that COBE be redesigned. (NASA Release 87-1)

• President Ronald Reagan's proposed fiscal year 1987 budget called for a 12.7 percent increase in NASA funding. The intent of the increase in the NASA budget was to resume the Space Shuttle flights, which had been suspended after the Challenger disaster. The President's budget also called for nearly $767 million for work on the crew-tended Space Shuttle. No funds were included for the untended rockets program, despite recommendations by several expert panels that NASA was overdependent on the Space Shuttle. (*NY Times,* Jan 6/87)

January 9: NASA named the five astronauts selected as the crew of the first U.S. space flight since the Challenger accident. Frederick H. Hauck was selected to command the crew, which would include Air Force Colonel Richard O. Covey, the pilot; and mission specialists John M. Lounge, George D. Nelson, and Marine Major David C. Hilmers. For the first time, the crew would be astronauts who had flown on Shuttle missions. Frederick H. Hauck, named acting Associate Administrator of NASA Headquarters Office of

External Affairs in August 1986, led the successful 1984 Shuttle recovery mission that returned two broken communications satellites to Earth. (*Star-News*, Jan 10/87)

January 12: Representative Edward Boland (Democrat-Massachusetts) and Senator Jake Garn (Republican-Utah), who hold leadership positions on House and Senate appropriations panels with NASA oversight, recommended that NASA change the Space Shuttle program launch schedule. The two members of Congress asked NASA to move the ESA Ulysses mission, scheduled for launch sometime in 1989-90, to a Titan 4 launch in 1991. This change would permit U.S. planetary satellites Galileo, Magellan, and the Mars Observer to be launched in the 1989-90 period. NASA officials agreed to give serious consideration to the change, but expressed fear that another delay in the Ulysses launch, originally scheduled for 1983, might damage NASA's relationship with the ESA. (*AvWk*, Jan 12/87)

January 14: A National Research Council panel, set up to monitor the Shuttle redesign, informed Administrator James C. Fletcher that NASA would not be able to make safety changes in time to meet the February 1988 target date for the resumption of Shuttle flights. The panel, headed by retired Air Force General Alton D. Slay, praised NASA's top-to-bottom review of hundreds of items and procedures that could cause another disaster. It added, however, that several analyses of possible equipment failure and of items considered critical to safety would not be completed until the summer of 1987. The panel recommended that NASA rank the most critical safety items, based on the likelihood of their failure, and that it closely link the engineering changes and the hazard analyses. (*NY Times*, Jan 15/87; *W Post*, Jan 16/87; *UPI*, Jan 16/87)

January 15: Dr. Lew Allen, Director of NASA's Jet Propulsion Laboratory (JPL), announced that construction of JPL's new Microdevices Laboratory would start January 21, 1987. On the same day, NASA Administrator Dr. James C. Fletcher and Dr. Marvin Goldberger, President of the California Institute of Technology, signed a memorandum of understanding for a Center for Space Microelectronics Technology to be established at JPL. The Center would be the successor to the Advanced Microelectronics Program established at JPL in 1983 and headed by Dr. Carl Kukkonen. The Center for Space Microelectronics Technology will be governed by a Board of Governors chaired by Dr. Lew Allen. Dr. Carl Kukkonen was named Center Director. (NASA Release 87-3; *Star-News*, Jan 20/87; *The Foothill Leader*, Jan 21/87)

January 16: NASA announced plans to launch two around-the-world balloon flights to examine newly discovered high energy x-ray microflares and flare plasmas being emitted by the Sun. The helium-filled, 28 million-cubic-foot-volume balloons are expected to be launched from Alice Springs, Australia, during January and February 1987. Taller than the Washington Monument, and carrying payloads weighing 3,000 pounds, the balloons are expected to reach an altitude of 130,000 feet. They would circle the Earth in 12 to 18 days and return to their launch site. (NASA Release 87-4)

January 20: NASA's Associate Administrator for Space Flight, Rear Admiral Richard H. Truly, announced that more than 400 modifications were planned for the Space Shuttle program. Speaking at a news conference called to discuss NASA's recovery from the effects of the Challenger accident, which killed all seven Shuttle crew members, Truly emphasized the Agency's commitment to "a true enhancement to safety." Included among the safety modifications listed were better brakes, more efficient engines, and eventually an escape system for the Shuttle crew. However, it was pointed out that the escape system probably would not be ready for the next planned Shuttle flight in February 1988. (*UPI,* Jan 20/87; *LA Times,* Jan 21/87; *NY Times,* Jan 21/87; *P Inq,* Jan 21/87; *W Post,* Jan 21/87)

January 21: The Air Force announced that it would purchase 20 rockets from the McDonnell Douglas Corporation for launching navigational satellites. Modeled on the existing rockets used by NASA, the Delta 2 medium rocket boosters would also be used commercially for launching communications satellites. (*UPI,* Jan 21/87; *USA Today,* Jan 22/87; *NY Times,* Jan 22/87; *W Post,* 22/87)

January 22: A number of senators expressed reservations that Morton Thiokol, Inc. remained the sole maker of NASA's Shuttle booster rocket. Members of the Senate Subcommittee on Science, Technology, and Space complained during the hearing that, despite the fact that Congress had instructed NASA to employ other industries to redesign Shuttle rockets, the Agency appeared locked into using rockets designed by Morton Thiokol, Inc. (*NY Times,* Jan 23/87; *WSJ,* Jan 23/87)

January 28: The first anniversary of the Challenger disaster was remembered by the Nation with a series of memorial services and ceremonies. Speaking at a Challenger accident commemoration, NASA Administrator Dr. James C. Fletcher stated that the terrible tragedy only reaffirmed NASA's "commitment to move the Nation forward into a new era of space flight, one more stable, more reliable and safer than before." The Challenger crew, he added, "would be pleased to know that the NASA family and the Nation are carrying on in that spirit." President Ronald Reagan, in a satellite message to NASA Centers, also saluted the seven "magnificent" Challenger astronauts. (*NY Times,* Jan 29/87; *W Post,* Jan 29/87; *W Times,* Jan 29/87)

January 29: NASA Administrator Dr. James C. Fletcher and General Kabbaj, the Inspector of the Royal Moroccan Air Force, signed an agreement permitting the United States the use of Morocco's Ben Guerir Air Force Base for emergency landing of the Space Shuttle. The facility would be ready for such an emergency in February 1988. (NASA Release 87-6)

• NASA announced that its numerical aerodynamic simulation (NAS) computer system would become operational in early March. Considered the most powerful computing system in the world, the NAS would help ensure continued national preeminence in aeronautical research. According to NASA, the research programs in which NAS would play a crucial role include work on the National Aero-Space Plane Program and a joint Department of Defense/NASA program in aerospace vehicle technologies and capabilities, including horizontal takeoff and landing. Development of new technology and validation program in this field might make possible a wide variety of operational aerospace vehicles, ranging from space launch vehicles to long-range air defense interceptors and hypersonic transports.

NASA emphasized that NAS is not a set of computer hardware, but an evolving capability. It is to be an array of 250 off-site scientists and engineers at 27 locations, accessing the system via satellite or high speed terrestrial lines. It is to be driven by the Cray 2 supercomputer, which has an enormous 256-million-word internal memory, 16 times larger than those of previous computers. (NASA Release 87-7)

• China signed a contract with Teresat, Inc. of New York to launch its first satellite for an American firm. The contract called for a Teresat Westar-VIS satellite to be launched into orbit aboard a Chinese-made rocket in the first half of 1988. (*LA Times,* Jan 29/87).

During January: The American journal *Aviation Week & Space Technology* reported two serious Soviet setbacks in space that were not acknowledged by Soviet authorities. The magazine claimed that on January 29 the Soviets deliberately destroyed a Cosmos 1,813 military reconnaissance satellite they had launched on January 15, 1987. When the satellite malfunctioned upon reentry, its self-destruction system was activated to prevent it from striking a populated area or from falling into western hands.

Also according to this source, a Soviet SL-12 Proton booster rocket failed during launching on January 30, 1987. Its fourth and uppermost stage failed to ignite and the debris from the rocket and communications satellite it carried fell out of orbit a day later. (*C Trib,* Feb 7/87; *NY Times,* Feb 7/87)

February

February 3: NASA issued a request for proposal to U.S. industry for a program support contractor to assist the NASA Space Station Program Office, in Washington, D.C., with development of the Space Station. (NASA Release 87-8)

• NASA officials declared that Space Station deployment, scheduled for 1994, could be delayed up to two years or the Station could be scaled down. NASA Administrator James C. Fletcher said that the cost of the Space Station, once estimated at $8 billion, was now estimated at between $13 and $14.5 billion. Fletcher, who said that the United States was lagging behind the Soviet Union in human space flight, predicted that a U.S. crew-tended Space Station would help the country catch up to the Soviets. Testifying before the Senate Commerce Subcommittee on Science, Technology and Space, Dr. Fletcher stressed that the Space Station program retained high priority with the Reagan administration. (*LA Times,* Feb 4/87; *USA Today,* Feb 4/87; *NY Times,* Feb 5/8)

February 4: NASA's Jet Propulsion Laboratory (JPL) selected the Bendix Field Engineering Corporation, in Columbia, Maryland, for negotiations on a contract for continuation of maintenance and operations of the NASA Deep Space Network (DSN). The DSN provides tracking and data acquisition for planetary space flight missions and for NASA's crew-tended and untended space programs. The five-year contract, with an option for an additional five years, was expected to begin April 2, 1987. (NASA Release 87-9)

February 5: Jesse W. Moore resigned from NASA, effective February 8, 1987. Moore joined NASA in 1966. Among the varied positions he held included Deputy Director of the Solar Terrestrial Division, Office of Space Science (1978–1979); Director of Space Flight Division (1979–1981); Director of the Earth and Planetary Exploration Division (1981–1983); Acting Associate Administrator of Space Flight (April 1984–August 1984); and Associate Administrator of Space Flight (August 1984–January 1986). He was named Director of the Johnson Space Flight Center in Houston, Texas, in January 1986, and in October 1986 he moved to Headquarters as a Special Assistant to the Associate Administrator for Policy and Planning. (NASA Release 87-10; *NY Times,* Feb 6/87; *W Post,* Feb 6/87)

• NASA announced the realignment of external communications functions within the Office of External Relations and the new Office of Communications. Both would be headed by an associate administrator. The realignment was in response to a NASA management study conducted by General Samuel C. Phillips. (NASA Release 87-11)

February 6: Two Soviet astronauts were launched aboard the new space vehicle, Soyuz TM-2, toward a rendezvous with the orbiting Mir space station. This was the first crew-tended Soviet space launch since July 1986 and only the second launch in the Russian human space program to be shown live on Soviet television. Western experts speculated that the two astronauts would aim for a long duration flight of at least six months. Mission commander Yuri Romanenko and flight engineer Alexander Laveikin were expected to begin link up operations with the Mir space station on February 7. (*NY Times,* Feb 6/87; *W Post,* Feb 6/87; *P Inq,* Feb 6/87; *C Trib,* Feb 7/87)

February 8: The *Christian Science Monitor* cited Soviet news agency TASS as reporting that the two Soviet cosmonauts aboard Soyuz TM-2 had successfully linked up with the Soviet space station Mir. (*CSM,* Feb 9/87)

February 11: The Air Force, after two consecutive failures in 18 months, successfully launched a Titan 3B rocket from Vandenberg Air Force Base in California. The launch placed a classified payload into a polar orbit around the Earth. Air Force Secretary Edward C. Aldridge hailed the launch as "the first major step" in the recovery of the American space program after the Challenger accident in January 1986. (*LA Times,* Feb 13/87; *NY Times,* Feb 13/87; *W Post,* Feb 13/87)

February 12: The United States and potential foreign partners agreed to resume negotiations aimed at building a Space Station in the 1990s. After a two-day meeting between representatives from the United States, Canada, Japan, and the 11-nation European Space Agency, agreement was reached that the project would remain a civilian endeavor. NASA negotiators emphasized, however, that they saw no conflict between the "peaceful purposes" of Space Station and potential use by the Pentagon for "nonaggressive" activities. (*NY Times,* Feb 13/87; *USA Today,* Feb 13/87)

February 17: NASA and the National Oceanic and Atmospheric Administration (NOAA) announced that a Geostationary Operational Environmental Satellite (GOES-H) would be launched sometime after February 23, 1987. The satellite would provide cloud cover images and atmospheric temperature profiles, collect space environment data, and conduct an experiment for detecting emergency distress signals on the ground from a geosynchronous orbit. (NASA Release 87-14)

February 19: Astronomer Carl Sagan told the Senate Commerce committee that a U.S. Mars mission would be the best cure for USA's "sagging" space program. According to Sagan, some 20,000 U.S. scientists have written to Congress urging a Mars mission. He recommended that NASA consider a possible joint mission with the Soviet Union, which had stepped up efforts to land on Mars. (*NY Times*, Feb 20/87; *USA Today*, Feb 20/87)

February 20: Robert Lovell and Fred Flatow, NASA search and rescue mission managers, were awarded the Yuri Gagarin medal by a space delegation from the Soviet Union. The awards were in recognition of their outstanding contribution to COSPAS/SARSAT, an international search and rescue program that uses satellites to locate people in distress. (NASA Release 87-17)

February 24: Dr. Leonard A. Fisk, Vice President for Research and Financial Affairs at the University of New Hampshire, was named NASA Associate Administrator for Space Science and Applications, effective April 6, 1987. He replaced Dr. Burton I. Edelson. (NASA Release 87-18)

• NASA announced that Morton Thiokol, Inc., the builder of the booster rocket that caused the Space Shuttle Challenger explosion, voluntarily accepted a $10 million reduction in fees and agreed to take no profit from $409 million worth of work required to fix future rockets. The company also agreed to replace the rockets lost in the accident. The agreement between Thiokol and NASA negated any lawsuits growing out of the January 28, 1986, accident that destroyed the Shuttle and killed the crew of seven. (*C Trib*, Feb 25/87; *W Post*, Feb 25/87; *B Sun*, Feb 25/87; *WSJ*, Feb 25/87)

February 25: A Soviet space vehicle, Progress-27, launched in January 1987 to bring supplies to the Mir space station, disintegrated as it reentered the Earth's upper atmosphere. (*P Inq*, Feb 27/87)

February 26: NASA launched a GOES-7 weather satellite aboard a Delta rocket. The satellite was first placed into an elliptical orbit. Ground command was expected to arrest the satellite in a stationary orbit 22,400 miles above the Atlantic Ocean by firing a booster rocket. GOES-7 joined the already orbiting GOES-6 weather satellite to provide weather information and patterns in an area stretching from the eastern Atlantic to the mid-Pacific Ocean. (*B Sun*, Feb 27/87; *P Inq*, Feb 27/87; *W Times*, Feb 27/87)

February 27: NASA reported that an 18-inch telescope, aboard a nine-year-old orbiting International Ultraviolet Explorer (IUE) satellite, was monitoring the intense emissions of ultraviolet radiation from recently discovered supernova in the Large Magellanic Cloud, a neighbor galaxy of our own Milky Way. (NASA Release 87-20)

During February: NASA's latest space technology led to new discoveries about ancient civilizations. Using Earth Observation Satellite and Earth Resources Observation Systems imagery and the Landsat 5 satellite, which has an advanced multispectral sensor Thematic Mapper, NASA's Ames Research Center scientists made first attempts to study Maya civilization settlement patterns, environmental settings, and natural resource use. (NASA Release 87-15)

March

March 1: NASA and the ESA announced solicitation of scientific investigation from the United States, European Space Agency member states, and Canada, for two missions. The four spacecraft CLUSTER mission would study basic plasma processes in the Earth's magnetosphere. The Solar and Heliospheric Observatory (SOHO) mission consisted of a spacecraft studying solar processes and solar-terrestrial relationships. (NASA Release 87-22)

March 4: NASA declared the Pioneer 9 spacecraft officially dead after a final attempt to contact it failed. The last signal was received on May 18, 1983. Pioneer 9 had orbited the Sun for 18 years. (NASA Release 87-23; *W Times,* Mar 5/87)

March 5: Willis H. Shapley was selected for NASA's newly established position of Associate Deputy Administrator (Policy), effective March 9, 1987. Shapley held the third ranking position, created in response to the reorganization recommended by the NASA Management Study Group.

Shapley began Government service in 1942 with the Bureau of the Budget, Executive Office of the President, where he specialized in research and development, national defense, and space programs. From 1965 to 1975, he was NASA's Associate Deputy Administrator. In July 1975 he was the senior NASA representative in the Soviet Union during the joint US–USSR Apollo-Soyuz space mission. Shapley retired from NASA in 1975 to become a consultant to a number of private institutions and Government agencies, including NASA. (NASA Release 87-24)

• NASA announced that beginning March 14, 1987, the Pioneer 12 space-craft, orbiting Planet Venus, would study the newly discovered Comet Wilson. The Comet was to be tracked by controllers at NASA's Ames Research Center, Mountain View, California, from March 14–21, and again from March 31 to April 30. (NASA Release 87-25)

March 6: James Elliott, a spokesman for NASA's Goddard Space Flight Center, Greenbelt, Maryland, announced that a NASA rocket exploded about 20 seconds after launch from Poker Flats, Alaska. The small rocket, carrying an upper-atmosphere plasma experiment, detonated when its second stage malfunctioned. (*W Post,* Mar 8/87)

March 9: NASA announced that its Numerical Aerodynamic Simulation (NAS) supercomputer system, the world's most advanced, had become fully operational. The NAS system will make possible a wide range of aerospace vehicle technologies and capabilities and will ensure the Nation's preeminence in aeronautical research. By the end of 1987, the operational system was expected to be capable of a billion calculations per second, and within a decade its capability should reach 10 billion calculations per second. The 90,000-square-foot NAS building, located at NASA's Ames Research Center, Mountain View, California, and containing a 15,000-square-foot central computer room was equipped with an array of systems for the optimal functioning of the computers. (NASA Release 87-21; *NY Times,* Mar 10/87; *CSM,* Mar 10/87; *SF Chron,* Mar 10/87)

• NASA Administrator Dr. James C. Fletcher announced that he would retain and strengthen the existing Space Shuttle processing arrangement at the Kennedy Space Center, in Florida, with Lockheed Space Operations Company, Titusville, Florida. The decision was recommended by a senior review group set up to study the ground processing of Shuttle flight hardware. The group was chaired by Roy S. Estess, Deputy Director, National Space Technology Laboratories, Mississippi, and included representatives from NASA Headquarters and from each field center. (NASA Release 87-26)

• *Aviation Week & Space Technology* reported that the Space Goals Task Force of the NASA Advisory Council recommended a crew- tended mission to Mars. The task force, headed by former Apollo 11 astronaut Michael Collins, stressed that the development and operation of a U.S./international Space Station was a prerequisite for exploration of Mars and beyond. It also stressed that NASA needed to resume Shuttle flights as soon as possible and develop expendable launch vehicles.

The task force listed the following major steps required for a successful Mars mission: an aggressive exploration of Mars to support a longer-term, crew-tended mission to Mars, buildup of the technological base needed to support that mission to Mars, establishment of a realistic schedule for the project, and an assessment of whether another crewed mission to the Moon was needed to precede a Mars mission. (*AvWk,* Mar 9/87; *NY Times,* Mar 18/87)

March 10: NASA announced formation of a group of scientists to examine potential Space Station activities, with the aim of reducing the time between experiment concept development and publishable results. Chaired by Dr. David C. Black, NASA Headquarters Space Station chief scientist, the group was made up of scientists and researchers from NASA, the National Science Foundation, the National Institutes of Health, universities, and the private sector. (NASA Release 87-27)

March 11: NASA selected General Electric Company, Space Systems Division, Philadelphia, Pennsylvania, and Harris Corporation, Government Communication Systems Division, Melbourne, Florida, to negotiate contracts for Phase I system design studies of the second Tracking and Data Relay Satellite System (TDRSS) ground terminal. (NASA Release 87-29)

March 12: The NASA Advisory Council recommended that the space agency acquire a diversified fleet of expendable launch vehicles (ELVs) to preserve the Space Shuttle for missions requiring unique capabilities. (NASA Release 87-30)

• A spokesman at the NASA Jet Propulsion Laboratory said that NASA's Deep Space Network station near Canberra, Australia, configured with Australia's Parkes Radio Observatory, was observing radio wave emissions from Supernova 1987a, first detected by astronomers in the large Magellanic cloud on February 24, 1987. (NASA Release 87-31)

March 13: NASA announced that its fiscal year 1988 budget request did not include funding for the launching of the Mars Observer mission in 1990. Under pressure from the scientific community, NASA considered spending about $130 million for the launching of the Mars Observer from an Air Force Titan rocket. The Agency said it planned to launch the spacecraft in 1992 from the Space Shuttle. (NASA Release 87-32; *LA Times,* Mar 14/87; *NY Times,* Mar 15/87)

• NASA scientists successfully ordered the Voyager 2 space probe, which was some 2 billion miles away from Earth, to fire its thrusters changing the spacecraft's course for its 1989 encounter with the planet Neptune. The course was altered so that Voyager 2, launched on August 20, 1977, would avoid the debris that might be orbiting the planet. The spacecraft was expected to arrive on August 24, 1989, within 3,100 miles of cloud tops above Neptune's north pole and was expected to pass within 25,000 miles of Neptune's moon Triton. (*B Sun,* Mar 14/87; *P Inq,* Mar 15/87)

March 14: Pioneer 12, which had been orbiting the planet Venus, was directed, by controllers at NASA's Ames Research Center, Mountain View, California, to begin tracking the newly discovered Comet Wilson. This would be the fourth time that Pioneer 2, orbiting Venus since 1986, interrupted observations of the planet to study a comet passing Venus on its way to the Sun. (*LA Times,* Mar 15/87)

March 17: NASA began a three-day ground system test of the scientific instruments to be carried on the Hubble Space Telescope (HST). The five scientific instruments to be carried aboard the HST were the wide-field and planetary camera, the high resolution spectrograph, the faint object spectrograph, the high speed photometer, and a faint object camera. (NASA Release 87-35)

• NASA announced that it was studying four major initiatives to help determine the next major goal in space. The initiatives emerged from proposals by NASA's Strategic Planning Group and by a group headed by astronaut Dr. Sally Ride. The four initiatives were a considerably expanded study of Earth systems, an enhanced program of solar system exploration, establishment of a permanent scientific base on the moon, and human exploration of Mars preceded by untended missions. At the same time, the independent NASA Advisory Council recommended that exploration of Mars be the Nation's primary goal in space. (NASA Release 87-36)

March 18: NASA announced that it would begin, in late March, a new aircraft flight research program at Ames-Dryden Flight Research Facility, in Edwards, California. The aim of the program was the study of super maneuverability and prevention of dangerous spins and related crashes. (NASA Release 87-37)

• NASA announced that scientists and researchers from the Jet Propulsion Laboratory, Pasadena, California, would conduct, on March 23–26, the first in a series of tests related to development of the mobile satellite communications system. The test in Erie, Colorado, would be part of the Mobile Satellite Experiment program. (NASA Release 87-38)

March 19: NASA selected Computer Sciences Corporation, Silver Spring, Maryland, to negotiate a contract for systems, engineering, and support services. The services provided by the contract would be in support of the Mission Operations and Data Systems Directorate, Goddard Space Flight Center, Greenbelt, Maryland. (NASA Release 87-39)

March 20: NASA announced that it would launch a seventh Department of Defense Fleet Satellite Communications spacecraft from the Cape Canaveral Air Force Station, in Florida, no earlier than March 26, 1987. (NASA Release 87-40)

• NASA launched a Palapa-B2P, an Indonesian communications satellite, on a Delta 182 from Launch Complex 17B, Eastern Space and Missile Center, Cape Canaveral Air Force Station, in Florida. The satellite, intended to serve Indonesia's 13,677 islands, was placed into an elliptical orbit ranging from about 115 miles to 23,000 miles above the Earth. The Palapa-B2P satellite, built by Hughes Aircraft Company, was to be used as a backup for a failing relay station already in orbit. (*NY Times,* Mar 21/87; *B Sun,* Mar 21/87; *P Inq,* Mar 21/87; *NY Times,* Mar 22/87)

March 23: NASA Administrator Dr. James C. Fletcher and Chairman of France's Centre National d'Etudes Spatiales Jacques-Louis Lions signed a memorandum of understanding for a joint satellite oceanographic mission. The mission, called TOPEX/POSEIDON, which would chart ocean topography and observe ocean circulation on a global scale, was scheduled for a late 1991 launch. The TOPEX/POSEIDON mission would be part of the international World Climate Research Program. (NASA Release 87-41)

March 26: NASA selected Lockheed Engineering and Management Services Company to negotiate an engineering support contract for Johnson Space Center, Houston, Texas. The contract would consolidate work currently performed by Lockheed, McDonnell Douglas Corporation, and Northrop Services, Inc. (NASA Release 87-42; *WSJ,* Mar 27/87; *H Chron,* Mar 27/87)

• An Atlas-Centaur rocket, launched during a driving thunderstorm from Cape Canaveral, Florida, veered out of control 52 seconds after liftoff and was blown up by NASA safety officials. The launch was NASA's third major mission in 1987. Investigators trying to determine the cause of the accident said that data radioed to the ground from the rocket seconds before it careened off-course, which indicated a major electrical disturbance aboard. Investigators speculated that lightning, prevalent in the area at the time, may have caused the failure. (*W Post,* Mar 27/87; *W Post,* Mar 31/87; *W Times,* Mar 31/87)

March 27: NASA announced that Johnson Space Center in Houston, Texas, and Jet Propulsion Laboratory in Pasadena, California, were studying concepts for designing and building a robot space vehicle for Mars exploration and sample-return mission. If approved, the robotic space vehicle mission would be launched in 1998 and return with martian samples in 2001. The mission would precede human exploration of Mars. (NASA Release 87-44)

• Rear Admiral Richard H. Truly, NASA Associate Administrator for Space Flight, announced the composition of the Atlas Centaur 67 Investigation Board. The Board, chaired by Jon R. Busse, Director, Office of Flight Assurance, Goddard Space Flight Center, Greenbelt, Maryland, would investigate the loss of the Atlas Centaur 67 mission on March 26, 1987 and recommend corrective action. The mission vehicle was destroyed 51 seconds after an apparently normal liftoff. (NASA Release 87-45)

March 31: The Soviet Union launched a 20-ton, self-contained astrophysical observatory whose mission was to dock with the orbiting Mir Space Station. Viktor Blagov, the deputy mission director, said that the observatory would "open a new chapter" in human space flight and would become a key building block of a permanently habitable scientific space complex. When docked with the Mir Space Station, the 19-foot-long observatory, with a 13.6 foot diameter, would double the working space available and would create a sort of a "high rise" structure in space. (*P Inq,* Apr 1/87; *W Times,* Apr 1/87)

During March: India failed to launch an augmented satellite launch vehicle, developed by the Indian Space Research Organization, when the vehicle malfunctioned and fell into the Indian Ocean. The launch from Sriharikota, India, was the first of four such planned launches, and was to have placed a stretched Rohini series scientific satellite into a low-Earth orbit. (*AvWk,* Mar 30/87)

April

April 1: NASA selected five American principal investigators and five Italian scientists for a joint U.S.-Italian Tethered Satellite System (TSS-1) project. The TSS-1, a deplorable subsatellite to be "tethered" to the Space Shuttle via a retractable cable, was expected to lift aboard Columbia in the fall of 1990. (NASA Release 87-46)

• NASA Administrator Dr. James C. Fletcher announced NASA planned to develop an advanced solid rocket motor for the Space Shuttle (NASA Release 87-47)

April 3: NASA announced that Jet Propulsion Laboratory (JPL) Director Dr. Lew Allen selected a Virginia site for a Space Station Program office. The 110,000 square feet of leased space in Reston, Virginia, would house NASA's Space Station Program director and some 400 JPL and NASA personnel. Initial occupancy was planned for summer 1987. (NASA Release 87-49)

• NASA announced that it was proceeding toward design and development of a Space Station that would establish a permanent U.S. human presence in space by the mid-1990's. It would seek proposals from industry for a phased development of the Space Station, including estimates for an enhanced-capability Space Station configuration.

NASA reported that the first phase of the Space Station development, the revised baseline configuration, which would provide an initial, permanently habitable research capability by 1996, had been approved by the President. It would include a laboratory and habitation modules, four resource nodes, a polar orbiting platform, and experiment provisions outside of the pressurized modules. (NASA Release 87-50; *LA Times,* Apr 4/87)

• NASA Administrator Dr. James C. Fletcher and European Space Agency (ESA) Director General Reimar Luest announced that the Space Shuttle would be used for the launch of the joint NASA/ESA Ulysses mission to the Sun in October 1990. At the same time, they announced that the launch, also from the Space Shuttle, of the Galileo mission to Jupiter had been moved up to October/November 1989. Since the distance to the Sun is considerably shorter than to Jupiter, the Ulysses spacecraft would begin to transmit data in 1994, a year earlier than the Galileo spacecraft. (NASA Release 87-51)

April 6: NASA selected Boeing Computer Services Company, Seattle, Washington, for negotiating a technical and management information system contract in support of the Space Station program. (NASA Release 87-52)

April 9: NASA's Office of Space Science and Applications, at Headquarters in Washington, D.C., selected 29 life sciences investigators for space flight study. According to Dr. Arnauld E. Nicogossian, Director of NASA's Life Sciences Division, the first phase of the study, the definition phase, which is expected to begin in the early fall of 1987 and last less than a year, would be followed by selection of investigators for actual flights aboard the Space Shuttle. Chosen investigators could participate in future Space Shuttle missions anytime between 1991 and 1994, or even later. (NASA Release 87-54)

• Despite the efforts of Soviet cosmonauts Yuri Romanenko and Aleksander Laveikin, who had tended the space station since February 7, 1987, the first attempts to dock the Soviet Union's Kvant space module with the space station Mir failed. The Kvant, a self-contained astronomical observatory, was to be the first experimental module to link with Mir. (*NY Times,* Apr 10/87; *W Times,* Apr 9/87)

April 10: NASA signed an agreement with the General Dynamics Space Systems Division, San Diego, California, granting authority to use NASA-controlled facilities and capabilities for commercial manufacture and launch of the Atlas/Centaur vehicle. This was the first instance of the U.S. Government transferring commercial operations of an expendable launch vehicle to the private sector. The agreement reflected NASA's endorsement of U.S. policy calling for the development of private sector launch capabilities. (NASA Release 87-55)

April 13: Soviet cosmonauts Yuri Romanenko and Aleksander Laveikin left their space station and successfully tightened the seal between the research module Kvant and the space station. Kvant had failed to dock on two previous tries. (*NY Times,* Apr 13/87)

April 15: John W. Young, chief of the astronaut office since January 1974, was appointed as Special Assistant to the Director for Engineering, Operations, and Safety at the Johnson Space Center (JSC), Houston, Texas. Young was expected primarily to advise the Director and other senior managers at JSC on issues affecting the safe return to flight of the Space Shuttle. According to JSC Director Aaron Cohen, "John Young's acceptance of this new responsibility will strengthen the link between operational and engineering elements at the Johnson Space Center." (NASA Release 87-58)

• The United States and the Soviet Union signed an accord on cooperation in space exploration. The accord culminated a three- day visit to Moscow by Secretary of State George P. Shultz. The accord formalized the ongoing exchange of information between NASA and the Soviet Academy of Sciences. It provided for no dramatic joint venture in space. Rather it called for coordination of projects in solar-system exploration, space astronomy and astrophysics, earth sciences, solar-terrestrial physics and space biology, and medicine. (*NY Times,* Apr 16/87; *P Inq,* Apr 16/87)

April 20: NASA announced plans for three balloon flights to study ray radiation produced by Supernova 1987a. The Goddard Space Flight Center's Wallops Flight Facility at Wallops Island, Virginia, which manages NASA's balloon program, would provide launch services and helium. The first launch, from Alice Springs, Australia, was to take place no earlier than May 7, 1987, and the third balloon was to be launched within a month's time. (NASA Release 87-60)

April 22: An international panel of scientists selected three experiments submitted by NASA's Ames Research Center, Mountain View, California, for the Comet Rendezvous Asteroid Flyby mission to be launched in early 1990's. The cometary ice and dust experiment (CIDEX) was designed to capture and analyze dust, ices, and gases evolved from the comet as the coma and tail develop during its pass around the sun. The thermal infrared radiometer experiment (TIREX) would measure the thermal emission and the scattering of solar radiation from the comet. The third experiment would consist of an interdisciplinary scientist for exobiology, the study of chemical processes and physical events leading to emergence of life in the universe. The CIDEX mission was managed by NASA's Jet Propulsion Laboratory in Pasadena, California. (NASA Release 87-63)

April 24: NASA reported that two techniques were being used to monitor the Supernova SN 1987a. In the first case, the Supernova was observed by NASA's Deep Space Network (DSN) located at Tidbinbilla, Australia, and connected by microwave with Australia's CSIRO Parkes Radio Telescope 200 miles away.

Dr. Robert Preston of NASA's Jet Propulsion Laboratory, which operates the DSN for NASA, said that the second technique, called Very Long Baseline Interferometry, consisted of an even wider network of four antennas: Tidbinbilla, Parkes, a Landsat ground station in central Australia; and a 26-meter dish at Hobart on Tasmania, an island southeast of Australia.

The supernova, an exploding star whose emissions were just arriving on Earth at the time of this report, was detected in the neighboring Large Magellanic Cloud galaxy. It was about 163,000 light years from the Earth. Supernova 1987a was the first such detectable star explosion close to our own galaxy since 1604. (NASA Release 87-64)

• NASA issued Requests for Proposals to U.S. industry for a detailed design and construction of a permanently habitable Space Station. The two options proposals were due by July 21, 1987. Option one was for a phased program calling for a permanently tended Space Station to be operational by 1996. Option two was the enhanced-capability Space Station configuration. (NASA Release 87-65; *LA Times*, Apr 25/87; *NY Times*, Apr 25/87)

April 27: NASA announced that its scientists had joined with scientists from the Brazilian Institute for Space Research and from U.S. and Brazilian universities and institutes to conduct a 45-day study of tropospheric (part of the atmosphere closest to Earth) chemistry in the atmosphere over the Amazon River Basin tropical rain forest during the wet season. NASA's program of global tropospheric experiments to study the chemistry of Earth's atmosphere and its interaction with land and oceans began in the 1980's. NASA used the Space Shuttle in 1981 and 1984 to measure tropospheric carbon monoxide with a gas filter radiometer. In July and August 1985, NASA's Amazon boundary layer experiment (ABLE) documented the importance of biosphere-atmosphere interactions in determining chemical processes in the troposphere over undisturbed rain forests. The second ABLE experiment, begun in mid-April 1987, was to study how tropical forests affect the gas exchange, chemistry, and budgets of several key gases in the troposphere.

A NASA Electra aircraft, stationed in Manasus, Brazil, in the center of the Amazon River Basin, was expected to measure atmospheric trace gases and aerosols during a series of flights. The results from the experiments aboard the Electra aircraft and from several investigations carried out at ground sites would be evaluated by the team of scientists late in 1987.

NASA's Office of Space Science and Applications in Washington, D.C., was responsible for the tropospheric chemistry program. Langley's Atmospheric Sciences Division, with logistical support from the Bionetics Corporation in Hampton, Virginia, managed the ABLE project.(Nasa Release 87-66)

• NASA announced that within hours of its discovery on February 24, 1987, NASA scientists were monitoring the Supernova 1987a with the Earth-orbiting International Ultraviolet Explorer (IUE) and the Solar Max Mission (SMM) spacecraft. Subsequently, as previously announced, NASA began monitoring the supernova with a network of ground antennas. The major supernova science program mounted by NASA and the intense, worldwide scientific interest in the supernova resulted from the fact that the discovery of Supernova 1987a provided scientists the opportunity to study not only the death of the star, but also the resutant rebirth of matter. The explosion, shock waves, and tremendous energies of Supernova 1987a were expected to provide direct evidence of the creation of heavy elements such as silicon and nickel.

This marks the first time a supernova has been so close to the Earth and so bright since the 1604 supernova, when the telescope had not yet been discovered. The advantageous circumstances allowed scientists to study the supernova in all radiation wavelengths from nearly the moment of its explosion. Scientists expected that the array of instruments and the variety of NASA's methods of monitoring the supernova would not only yield a new understanding of the chemistry and physics of the comet, but also potentially would provide a better understanding of the creation of matter.

The supernova program was expected to extend at least over the next 2-1/2 years. It would consist of multiple, continuous satellite observations, balloon and sounding rocket missions, aircraft flights, and radio observations. A special computer communications network would allow scientists from NASA, American and foreign universities, and other international scientific organizations to analyze the great amount of data generated by the program.

In addition to the IUE and SMM spacecraft, managed and operated by NASA's Goddard Space Flight Center in Greenbelt, Maryland, a Japanese Ginga (Astro-C) satellite was used to monitor the Supernova 1987a. The IUE satellite provided data from ultraviolet observations and monitored the overall brightness of the supernova. It also was expected to help define exactly which star in the Large Magellanic Cloud exploded creating the supernova. The SMM satellite, a sun-observing instrument, was used to monitor the expected gamma ray emissions from the supernova. The Japanese Ginga (Astro-C) satellite was used to monitor the supernova for x-ray emissions.

NASA's Gerard P. Kuiper Airborne Observatory (KAO), a C-141 jet transport stationed in New Zealand, was also used in supernova observations to measure infrared energy. It was expected to be involved in a number of flights through the spring of 1989, using new and increasingly more sophisticated instruments. (NASA Release 87-67)

April 31: Martin Marietta Corporation of Bethesda, Maryland, formed a new organization, the Martin Marietta Commercial Titan Systems, to sell commercial rockets. The new organization, to be headed by Richard E. Bracken, former head of Space Launch Systems at Martin Marietta in Denver, would sell a modified version of the Air Force Titan 3 rocket to owners of commercial satellites. (*W Post,* May 4/87)

During April: NASA and the Israel Space Agency signed an agreement allowing for the first Israeli space experiment on a future Space Shuttle flight. (NASA Release 87-57; *Science,* vol. 236, May 1/87)

Two top U.S. scientists, serving in advisory positions to NASA, resigned following their criticism of NASA policy and the space station. On April 10, 1987, Thomas Donahue, Chairman of the Space Science Board, National Academy of Sciences, resigned after sending a letter to some members of Congress calling for NASA to place higher priority on expendable launch vehicles for space science missions than on an as yet "poorly defined space station." Donahue was reinstated on April 15, 1987, after resolving his disagreement with Frank Press, head of the National Academy of Sciences.

The other scientist to resign was Peter Banks, Chairman of NASA's Task Force on the Scientific Uses of the Space Station. Banks resigned on April 13, 1987, after he criticized the Space Station planning, saying that the schedule for operations should be accelerated and a heavy-lift launch vehicle should be used. (*AvWk*, Apr 20/87)

May

May 6: NASA Assistant Administrator for Commercial Programs Isaac T. Gillam named 28 teams that have submitted proposals to establish Centers for the Commercial Development of Space. The objective of the Centers would be to stimulate research in the microgravity environment of space. This research was expected to lead the development of new products that have commercial value. The proposed research fields include space power, space propulsion, remote sensing, communication technology, and bioscience. The overwhelming majority of the proposals came from American universities. (NASA Release 87-69)

May 7: NASA announced that it, together with the Universities Space Research Association, in Columbia, Maryland, and the University of Maryland, College Park, Maryland, would establish a Center for Excellence in Space Data and Information Sciences at NASA's Goddard Space Flight Center in Greenbelt, Maryland, beginning October 1, 1987. The new research facility would serve as the nerve center for a network of researchers from leading universities and from industry computer science departments. The emphasis would be on research in areas of potential long-term application to NASA programs in the Space Station era. (NASA Release 87-70)

May 8: NASA announced that it had established a team to review launch decision processes used for NASA expendable launch vehicles and the Space Shuttle. The purpose of the review, which was to be completed by July 3, 1987, was to ensure consistent approaches and risk evaluations for all NASA launches. Joseph B. Mahon, Deputy Associate Administrator for Space Flight, and Robert L. Crippen, Deputy Director of the National Space Transportation System Operations, were selected as cochairmen of the team. (NASA Release 87-71)

May 10: Former astronaut Donald K. Slayton, President of Space Services, said that his company would be the first private U.S. company to launch satellites into orbit. According to Slayton, the first of five precision navigation-location Star Find satellites would be launched from NASA's Wallops Island in Virginia in 1988. Space Services would use its Conestoga II rocket system to launch the 300-pound satellite into a geostationary orbit 22,000 miles above the Earth. (*W Post,* May 11/87; *W Times,* May 11/87)

May 11: NASA released the findings of the Accident Investigation Board, which had looked into loss of the Atlas/Centaur-67 on March 26,1987. In a report to Rear Admiral Richard H. Truly, NASA Associate Administrator for Space Flight, Jon R. Busse, Chairman of the Accident Investigation Board, blamed the accident on an electrical transient, caused by a single-triggered lightning flash. Furthermore, the Board found that the launch was made "in violation of the established criteria used to avoid potential electrical hazards." The Board made recommendations concerning weather criteria, the launch decision process, and future launch vehicle. (NASA Release 87-72)

May 15: The Soviet Union launched a giant new rocket booster more powerful than the American Space Shuttle. According to the Soviet Tass News Agency, the first two stages of the rocket, launched from the Baikonur test site in Central Asia, fired correctly but a mock satellite failed to enter orbit and landed in the Pacific Ocean. Tass described the untended Energia launch vehicle as a "two-stage multipurpose launch vehicle" capable of putting into orbit more than 100 tons of payload, which is 75 tons more than what the U.S. Space Shuttle could carry. (*LA Times,* May 17/87; *C Trib,* May 18/87; *W Times,* May 18/87)

• In an effort to reduce reliance on the U.S. Space Shuttle, NASA would rely more heavily on untended rockets, declared Dr. James C. Fletcher, NASA Administrator. He said that the primary goal was to "accelerate the deployment of the Nation's backlog of space-science missions." NASA hoped to launch as many as five satellites and space probes by untended rockets by 1992. (*WSJ,* May 18/87)

May 19: Officials from NASA and the Lockheed-Georgia Company, Marietta, Georgia, announced that flight testing had begun on a new, highly fuel-efficient, "propfan" engine. Aircraft equipped with the propfan propulsion system, which was developed as a result of a $56-million NASA sponsored research program, was expected to be 15–30 percent more fuel efficient than the most advanced turbofan-powered aircraft flying in the 1990's. (NASA Release 87-73)

• The Soviet Union launched another cargo ship with supplies for the space station Mir. According to Tass, the cargo ship Progress 30, the third supply ship launched since cosmonauts Yuri Romanenko and Alexander Laveikin had occupied the space station in February 1987, was functioning normally after the launch. (*LA Times,* May 20/87)

May 20: NASA announced that the first post-Challenger Space Shuttle launch had again been rescheduled, this time for June 1988. The four-month delay was caused by the need to perform further tests on the liquid-fuel main engines and because of needed redesigning and testing of the Shuttle's rocket boosters.

The postponement caused a decrease in the number of crew-tended flights planned for 1988 and 1989. NASA said that three Space Shuttles, instead of the five originally planned, would be flown in 1988 and seven, instead of the ten planned, would be flown in 1989. The reductions in Space Shuttle launches, NASA believed, would further slow U.S. efforts to regain primacy in space exploration and utilization. (*LA Times*, May 21/87; *NY Times*, May 21/87; *B Sun*, May 21/87; *W Post*, May 21/87)

May 26: NASA announced that astronaut Dr. Sally K. Ride would leave the agency in Fall 1987. Dr. Ride, the first American woman to fly in space, would assume the position of Science Fellow at the Stanford University Center for International Security and Arms Control in Palo Alto, California. (NASA Release 87-84; *B Sun*, May 27/87; *NY Times*, May 27/87; *W Post*, May 27/87; *W Times*, May 27/87)

May 27: The first full-scale test of the redesigned Space Shuttle booster since the Challenger accident was labeled a spectacular success by all who watched it. The firing of the 126-foot Shuttle booster in Utah's Wasatch Mountains, witnessed by 300 officials from NASA and Morton Thiokol, Inc., the company that designed the booster rocket, was the first of six full-scale tests. Also present at this test were the five astronauts scheduled to fly the Space Shuttle in June 1988. (*B Sun*, May 28/87; *NY Times*, May 28/87; *W Post*, May 28/87)

May 28: NASA issued a request to the U.S. aerospace industry for system studies and design concepts of liquid-fueled rocket boosters. The request for proposals resurrects the original concept for liquid-fueled rockets for the Space Shuttle that NASA had rejected. The request for studies was part of NASA's efforts to find an alternative to the hard-to-control solid-fuel boosters. Four NASA Centers—Marshall Space Flight Center, Huntsville, Alabama; Johnson Space Center, Houston, Texas; Kennedy Space Center, Cape Canaveral, Florida; and Langley Research Center, Hampton, Virginia—would be involved in this project. (NASA Release 87-85; *H Chron*, May 29/87; *H Post*, May 29/87)

May 29: NASA Administrator Dr. James C. Fletcher announced the promotion of Dr. John M. Klineberg, Deputy Director of Lewis Research Center, Cleveland, Ohio, since July 1979, to Director of the Center, effective immediately. Dr. Klineberg went to Lewis from NASA Headquarters, where he had served as Deputy Associate Administrator for Aeronautics and Space Technology. An internationally recognized expert in the field of transonic flow, he first joined NASA in 1970 at the Ames Research Center, Mountain View, California. (NASA Release 87-86)

June

June 1: NASA announced the creation of an Office of Exploration, responsible for coordinating the Agency's missions to "expand the human presence beyond the Earth." Dr. Sally K. Ride, who had been in charge of a NASA study group to determine U.S. goals in space, was named the Acting Assistant Administrator until mid-August 1987. (NASA Release 87-87; *H Post,* June 2/87)

June 3: An Information Science Division was established at NASA's Ames Research Center, Mountain View, California, to conduct research in artificial intelligence for applications to space exploration. It was expected that artificial intelligence would be used to develop automated systems for all phases of space missions, from launch to mission control and for onboard operations. Artificial intelligence technology would also be used to advance spacecraft performance and safety, and would free astronauts from routine chores.

Research in artificial intelligence and the development of automated system would be a joint effort between the Information Science Division, other NASA Centers, private industry, and academic institutions. (NASA Release 87-89)

• NASA announced that Marshall Space Flight Center, Huntsville, Alabama, had invited industry to develop advanced solid rocket motor designs. The preliminary designs called for both a segmented motor design and a monolithic motor design. Requests for these designs, called Phase B studies, followed completion of "Phase A" conceptual studies of alternative solid rocket motor designs. Those designs were done by five aerospace firms. (NASA Release 87-90)

June 5: NASA announced that 5 civilians and 10 military officers were selected as new astronaut candidates for the Space Shuttle program. Of the 15, 7 were chosen as pilot astronaut candidates and 8 as mission specialist candidates. The elite group also included the first Black woman, Dr. Mae C. Jemison, to be chosen for the program. (NASA Release 87-93; *NY Times,* June 6/87; *LA Times,* June 6/87; *H Chron,* June 6/87)

June 8: According to a news release from NASA's Goddard Space Flight Center, Greenbelt, Maryland, NASA would manage a major international experiment on the use of satellites to study our environment. The experiment, part of the International Satellite Land Surface Climatology Project, would employ over 100 scientists from more than 25 institutions to study the interaction between the land system and climate system. Over 25 ground measurement stations would be set up on the Konza Prairie in Kansas for the experiment. Six research planes and five Earth-orbiting satellites would also be used. (NASA Release 87-95)

June 10: NASA announced that lightning had destroyed an Orion rocket scheduled for launch. The intense storm at the Wallops Flight Center, Wallops Island, Virginia, also destroyed the scientific experiment payload. (NASA Release 87-97; *NY Times,* June 11/87; *W Post,* June 11/87; *H Chron,* June 11/87)

June 12: In a major organizational shift, NASA announced the adoption of new post-Challenger era commercial activity policies. The released decision memoranda, covering the new policies, were signed by Deputy Administrator Dale D. Myers. The new policies resulted from the report of the Commercialization of Space Review Task Force and covered such general activities as joint endeavor agreements, microgravity research, secondary payloads on the Space Shuttle, expendable launch vehicles, and commercial activities.

According to the new policies, the Office of Commercial Programs would henceforth determine, in coordination with concerned Headquarters offices, whether NASA should participate in any proposed joint program. Responsibility for programs utilizing the microgravity environment was transferred from the Office of Aeronautics and Space Technology to the Office of Space Science and Applications. Secondary payload space would be carefully allocated by the Deputy Administrator and implemented by the Office of Space Flight to accomplish specific NASA objectives. To encourage and facilitate as much as possible the private domestic access to space, NASA's Office of Space Flight would contract private industry for launch of Expendable Launch Vehicles. Finally, the Office of Commercial Programs would, to the fullest extent possible, facilitate the utilization of space for commercial purposes. (NASA Decision Memorandum on Commercialization Policy, June 12/87; NASA Decision Memorandum on Responsibility for Programs Utilizing the Microgravity Environment, June 12/87; NASA Decision Memorandum on Secondary STS Payloads, June 12/87; NASA Decision Memorandum on Expendable Launch Vehicles (ELV's) Policy, June 12/87; NASA Decision Memorandum on Commercialization Policy, June 12/87; *LA Times,* June 13/87)

June 17: NASA announced reopening, after major modifications, the wind tunnel at Ames Research Center, Mountain View, California. Constructed in 1944, the 40-by-80-foot wind tunnel was used for full-scale testing of civilian and military aircraft. Modifications to the tunnel included increasing power from six 6,000-horsepower electric motors generating airspeeds of up to 230 mph to six 22,500-horsepower motors generating top airspeeds of 345 mph.

The modified 40-by-80-foot wind tunnel represented the first phase of the creation of the National Full-Scale Aerodynamics Complex, which included the Outdoor Aerodynamics Research Facility, at the Ames Research Center, Mountain View, California. The second phase, including the new "straight through" leg with its 80-by-120-foot test section, was expected to be completed in late 1987. (NASA Release 87-98)

• NASA announced two new management appointments effective June 22, 1987. Dr. Noel W. Hinners, Director of NASA's Goddard Space Flight Center, Greenbelt, Maryland, was appointed to the newly created position of NASA Associate Deputy Administrator (Institution). The position was created on recommendations of the NASA Management Study Group to provide focus on institutional management matters. NASA appointed Dr. John W. Townsend, Jr. to replace Dr. Hinners as Director of Goddard.

Dr. Hinners joined NASA in 1972 as director of Lunar Programs, in the Office of Space Science. From 1974 until 1979 he served as NASA Associate Administrator for Space Science. From 1979 to 1982, Dr. Hinners was Director of the Smithsonian Institute's National Air and Space Museum. In 1982, he was appointed Director of Goddard Space Flight Center.

Dr. Townsend, a research physicist at the Naval Research Laboratory, was transferred with his branch and the Vanguard Project to NASA in 1958 and became chief of the space science division. He became Assistant Director in 1959 and Deputy Director in 1965 of Space Science and Satellite Applications at Goddard Space Flight Center. In 1968 he was appointed Deputy Director of the Environmental Science Service Administration (ESSA), Department of Commerce. He became the Associate Administrator of ESSA, when ESSA became part of NASA in 1970. (NASA Release 87-99)

• Rear Admiral Richard H. Truly, NASA Associate Administrator for Space Flight, announced that the next Space Shuttle orbiter would be assembled in Palmdale, California. He gave costs and schedule factors as the reasons why this site was chosen over Califonia's Vandenberg Air Force Base (NASA Release 87-100; *H Chron,* June 18/87)

June 22: NASA announced that on June 15, 1987, Dr. John-David F. Bartoe began a 2-year appointment as Chief Scientist of the Space Station program, advising the Associate Administrator on program plans, policies, and user-related requirements. Prior to this appointment, Dr. Bartoe served with the Naval Research Laboratory (NRL) in a variety of positions, most recently as Head of the Solar Spectroscopy section, Space Science Division. He expected to return to the NRL at the end of the Chief Scientist appointment. (NASA Release 87-103)

June 29: NASA announced selection of the Astronautics Division of the Lockheed Missiles and Space Company, in Sunnyvale, California, for the design, development and implementation of the Software Support Environment (SSE) for the Space Station. The SSE contract was to be managed by the Space Station Program Office in Reston, Virginia. (NASA Release 87-107)

June 30: A report sent to President Ronald Reagan by NASA Administrator Dr. James C. Fletcher stated that the first post-Challenger Space Shuttle would have an escape hatch. But a decision on the rocket assist system necessary to eject the crew safely in emergency had not yet been made. The escape hatch was one of many significant changes made in the Shuttle Discovery, scheduled for launch in June 1988. Twenty-one major changes were made in the Shuttle's main engines alone, in order to increase operating life, safety, and reliability. These and other changes listed in the NASA report to the White House reflected recommendations made by the Rogers Commission, following their investigation of the Challenger accident. (*C Trib,* July 2/87; *LA Times,* July 1/87; *NY Times,* July 1/87)

During June: General Dynamics Corporation said that it would build 18 rockets over the next five years to launch communications satellites. According to a company spokesman, General Dynamics would spend $100 million to start production of the Atlas-Centaur rocket. General Dynamics became the third U.S. company, together with McDonnell Douglas Corporation and Martin Marietta Corporation, to attempt to move into launch vehicles production after the Challenger accident. (*WSJ,* June 16/87)

July

July 2: NASA granted the Space Station Program Support Division, of Grumman Aerospace Corporation in Bethpage, New York, a contract to provide systems engineering and integration for the Space Station. The contract also called for Grumman to provide a broad base of management support to the Space Station Program Office in the areas of program control and management; information systems; operations; program requirements and assessment; and safety, reliability, and quality assurance. (NASA Release 87-110; *WSJ*, July 3/87; *B Sun*, July 3/87)

July 13: NASA's last Atlas-Centaur rocket was extensively damaged in a launch pad accident. A $4-million fuel tank was destroyed when it was hit by a work platform. Four General Dynamics Space Division workers were treated for minor injuries. The launch of a military satellite, scheduled for liftoff aboard the Atlas-Centaur rocket in the fall of 1987, was grounded indefinitely. (*LA Times,* July 14/87)

July 15: NASA announced that it would showcase exhibits and models of its advanced fuel-saving propeller system called "propfan" at the 35th Annual Experiment Aircraft Association International Fly-In Convention and Sport Aviation Exhibition, July 31–August 7, in Oshkosh, Wisconsin. Propfan engines have revolutionary propellers with thin, highly swept blades. They burn 30 percent less fuel than advanced turbofan engines under the same conditions. NASA also was expected to exhibit its "Cockpit of the Future," in which pilots would use color-coded graphic displays and computers to monitor flight path and speed, wind and weather conditions, time-critical alerts, and traffic conditions and obstacles. Other displays would feature innovations in wing design; new composite materials to make lighter aircraft; new de-icing techniques and other safety improvements; a video presentation of NASA's Numerical Aerodynamic Simulation (NAS) supercomputer system; and models of Langley's National Transonic Facility, the largest cryogenic wind tunnel in the world, and the Ames Vertical Motion Simulator used to investigate landing, takeoff and other handling qualities, and to develop controls for short takeoff and vertical-landing aircraft. (NASA Release 87-111)

• NASA named a board to investigate the July 13, 1987, Atlas Centaur 68 accident which resulted in extensive damage to the vehicle. James B. Odom, Director of the Science and Engineering irectorate, Marshall Space Flight Center, Huntsville, Alabama, was appointed Chairman of the Board. The

Board was expected to report the results of its investigation to Rear Admiral Richard H. Truly, Associate Administrator for Space Flight, by August 14, 1987. (NASA Release 87-112; *LA Times,* July 14/87; *NY Times,* July 16/87)

July 16: NASA announced that it had chosen the final design for the Space Shuttle's new rockets. Also, the Agency declared it had resolved the problem of Challenger's leaking seals. Hot gasses escaped during launch and the seals failed to close properly because of cold weather. John Thomas, Manager of the rockets design team at the Marshall Space Flight Center in Huntsville, Alabama, stated that henceforth the o-ring seals would be heated in cold weather rather than constructed of new materials. He also said that the new design would contain three o-ring seals instead of two. Thomas emphasized that up to eight more test firings of the Space Shuttle's solid rocket motor were planned before flights resume. (*NY Times,* July 17/87; *W Post,* July 17/87; *W Times,* July 17/87)

July 21: Pilots from NASA, the U.S. Air Force, the U.S. Navy, and the Grumman Aerospace Corporation completed the first phase of experimental flying of the forward-swept-wing X-29 aircraft. During the tests, the X-29, with its three-surface pitch control system, reached an altitude of 50,200 feet and speeds of about 1.5 times the speed of sound.

After a total of 104 flights, the pilots began the second phase of the flight research program at NASA's Ames-Dryden Flight Research Facility in Edwards, California. During this phase, further flights would test the forward swept wing structure divergence tendencies and the overall aerodynamic performance of the wings and canards. Researchers would also investigate the aircraft's control system and handling qualities, as well as maneuvering characteristics applicable to military use. (NASA Release 87-113)

July 22: The Soviet Union launched into orbit a Soyuz TM-3 with two Soviet cosmonauts and a Syrian space traveler aboard. The liftoff from Baikonur Cosmodrome, in Soviet Kazakhstan, was the Soviet Union's third crew-tended space launch since the U.S. Space Shuttle explosion on January 28, 1986. The Soyuz TM-3, piloted by Alexander Viktorenko and Alexander Alexandrov, was scheduled to dock with the Soviet Mir space station after two days of orbiting around the Earth. The Syrian space traveler, Lieutenant Colonel Mohammed Faris, was the second Arab in space. The first was the Saudi Prince Sultan bin Salman, a nephew of Saudi Arabia's King Fahd. Prince Sultan bin Salman was a payload specialist aboard the U.S. Shuttle Discovery in 1985. (*H Chron,* July 22/87)

July 28: NASA selected seven new Centers for the Commercial Development of Space (CCDS). The Centers would conduct research leading to the development of new technologies that could be exploited commercially in space. The combined industry and university centers selected were the University of Tennessee Space Institute—Center for Advanced Space Propulsion, Tullahoma, Tennessee; Auburn University—Center for the Commercial Development of Space Power, Auburn, Alabama; Environmental Research Institute of Michigan—Center for the Commercial Development of Autonomous and Man-Controlled Robotic Sensing System in Space, Ann Arbor, Michigan; Pennsylvania State University—Center for Secretion Research, University Park, Pennsylvania; University of Colorado—Center for Bioserve Space Technologies, Boulder, Colorado; Case Western Reserve University—Center on Materials for Space Structures, Cleveland, Ohio; and Texas A&M Research Foundation—Center for Commercial Development of Space Power, College Station, Texas. (NASA Release 87-115)

• NASA announced an international joint effort to investigate ozone depletion over the Antarctic. NASA was to be joined in this effort by other Federal science agencies; the Chemical Manufacturers Association; scientists from Harvard University, the University of Denver, and the University of Washington; and the governments of Argentina, Chile, France, Great Britain, and New Zealand. A specially equipped ER-2 plane and a modified DC-8 airliner were to make a number of flights through the ozone hole between August 17 and September 29, 1987, to see whether man-made chemicals, nature, or both are destroying the ozone. The findings will be examined by an international panel of policy makers meeting in Montreal, Canada, in September 1987. (*NY Times,* July 30/87; *W Post,* July 29/87; *W Times,* July 29/87)

July 29: Top astronaut John W. Young told the *Orlando Sentinel* that he had been forced to resign as chief of NASA's astronaut office. Young claimed that his reassignment to a new post, as special assistant to the director for Shuttle engineering and safety, stemmed from his frequent criticism of NASA safety policies after the Challenger disaster. (*LA Times,* July 30/87)

July 30: Astronaut John A. McBride was named acting Assistant Administrator for Congressional Relations, effective September 1, 1987. He would replace John F. Murphy, who was appointed Chief of the Office of External Affairs at NASA's Ames Research Center, in Mountain View, California. (NASA Release 87-116)

July 31: Rear Admiral Richard H. Truly, Associate Administrator for Space Flight, announced NASA had completed negotiations with Rockwell International to build a replacement Space Shuttle orbiter. The new orbiter would contain existing structural spares, but would also feature the latest technology evolving from the current return-to-flight activities. The new Space Shuttle orbiter was expected to be ready for use within 45 months. (NASA Release 87-117)

• NASA said it had completed negotiations with Rockwell International Corporation for a new Space Shuttle. The new space vehicle, to replace the Challenger lost in the 1986 accident, was expected to be delivered in April 1991. The cost of the new Space Shuttle was estimated at $1.3 billion. (*C Trib,* Aug 2/87; *H Post,* Aug 1/87; *LA Times,* Aug 1/87; *B Sun,* Aug 1/87)

August

August 3: Five NASA and U.S. Air Force pilots completed the first phase of research flying of the Advanced Fighter Technology Integration F-111 aircraft at NASA's Ames-Dryden Flight Research Facility in Edwards, California. During the first phase of research flying, the aircraft's wing was modified so that it could fly with optimum wing curvature at subsonic, transonic, and supersonic speeds. The MAW system was operated only manually, and the pilot selected the aircraft's wing curvature. The aircraft, with its distinct Mission Adaptive Wing (MAW), entered the second phase of research flying where the computers were modified so the wing curvature could adjust automatically. (NASA Release 87-118; *Def News,* Aug 24/87)

August 5: NASA announced plans to study the feasibility of building an untended rocket to carry heavy materials for use in Space Station construction. The proposed concept calls for a space freighter capable of carrying more than three times as much cargo as the Space Shuttle. The vehicle would use the same type of solid fuel boosters as the Space Shuttle. NASA surmised that such a vehicle, whose estimated costs were put at $1.5 billion, could be delivered in 1993. (*H Chron,* Aug 6/87; *H Post,* Aug 6/87; *W Times,* Aug 6/87; *USA Today,* Aug 6/87)

August 7: Five aerospace firms were awarded contracts by NASA's Marshall Space Flight Center, Huntsville, Alabama, to develop preliminary advanced solid rocket motor designs for the Space Shuttle. The five firms receiving the awards were Aerojet Solid Propulsion Company, Sacramento, California; Atlantic Research Corporation of Alexandria, Virginia; Hercules Aerospace Company's Aerospace division of Salt Lake city, Utah; Morton Thiokol Inc., Brigham City, Utah; and United Technologies Chemical System Division, San Jose, California. The results of the preliminary studies would determine whether NASA will pursue the design, development, test and production of an advanced motor for Shuttle flights in fiscal year 1989. (NASA Release 87-120)

August 10: NASA announced that Dr. Noel Hinners would replace Dr. Frank McDonald as Chief Scientist, effective August 24, 1987. At the same time, Dr. Hinners would retain his position as NASA's Associate Deputy Administrator (Institutions). NASA also announced Dr. Frank McDonald would return to the Goddard Space Flight Center as Associate Director/Chief Scientist. (NASA Release 87-121)

• The 10-ton Soviet satellite Cosmos-1871 plunged harmlessly into the Pacific Ocean west of Hawaii. Soviet news agency Tass reported that the satellite, launched into a polar orbit on August 1, 1987, carried scientific equipment for space exploration. (*B Sun*, Aug 11/87; *W Post*, Aug 11/87; *W Times*, Aug 10/87)

• The Transportation Department announced that it has granted approval for Martin Marietta Corporation to launch private communications satellites. Within hours of the announcement, Martin Marietta signed an agreement with the International Telecommunications Satellite Organization to launch two Intelsat-6 satellites into orbit on Martin Marietta's Titan 3 rockets. Launchings were scheduled for 1989 and 1990. The agreement between two private companies to launch private communications satellites was the first instance of the implementation of the Reagan administration's space commercialization policy. (*B Sun*, Aug 11/87; *W Times*, Aug 11/87)

• American and Soviet scientists agreed to joint effort in a wide range of space research. Their first joint project called for a launch of a Soviet "space zoo" satellite containing monkeys, rats, fish, and other living organisms. The 14-day flight of this satellite was to provide Soviet and American scientists an opportunity to study the impact of weightlessness on the vital functions of animals. Some 26 other experiments were approved by working groups from the United States and the Soviet Union meeting for a week-long scientific conference in Moscow. The joint project was the first to be conducted under the terms of a five-year space cooperation agreement signed on April 15, 1987, by Soviet Foreign Minister Eduard A. Shevardnadze and U.S. Secretary of State George P. Shultz. (*C Trib*, Aug 11/87; *LA Times*, Aug 11/87; *P Inq*, Aug 11/87)

August 17: NASA released a long awaited report of a task force, headed by Dr. Sally K. Ride, set up to assess NASA's future space policy in the wake of the Challenger disaster. The 63-page report, delivered to NASA Administrator Dr. James C. Fletcher the preceding week, urged that the United States build a permanent outpost on the Moon as the first step toward the ultimate goal of "exploring, prospecting, and settling Mars." Ride emphasized, however, that while settling Mars should be "our eventual goal," it should not be "our next goal." She recommended a strategy of "natural progression which leads step by step, in orderly, unhurried way, inexorably toward Mars." (*LA Times*, Aug 18/87; *NY Times*, Aug 18/87; *W Post*, Aug 18/87; *W Times*, Aug 18/87)

• Flight controllers at NASA's Jet Propulsion Laboratory in Pasadena, California, reported that the untended spacecraft Voyager 2 was on course for a rendezvous in two years with the planet Neptune. Launched on August 20, 1977, Voyager 2, travelling at a speed of 41,600 miles-per-hour, was 642 million miles away from its target. It was expected to come within 3,000 miles of Neptune on August 24, 1989. (*NY Times*, Aug 18/87)

August 18: NASA announced that Morton Thiokol Space Division, a NASA contractor, would test fire the first redesigned Space Shuttle solid rocket motor on August 27, 1987. The 126-foot-long, 1.2-million-pound development motor (DM-8) would be fitted with 250 instruments to measure acceleration, pressure, deflection, thrust, strain, temperature, and electrical properties. (NASA Release 87-125)

August 19: NASA Administrator Dr. James C. Fletcher announced that Isaac T. Gillam, NASA Assistant Administrator for Commercial Programs, would retire effective September 1, 1987, and that Lawrence Herbolsheimer, Deputy Assistant Administrator in the Office of Commercial Programs, would serve as the Acting Assistant Administrator.

Gillam had served in a number of positions at NASA. In June 1978 he was appointed director of NASA's Dryden Flight Research Center in Edwards, California. Prior to that appointment, he was Dryden's Deputy Director and Director of Shuttle Operations. He subsequently served at NASA Headquarters as Special Assistant to the NASA Administrator and then moved on to head the Office of Commercial Programs. (NASA Release 87-126)

August 26: A solid rocket motor intended for use in NASA's Delta launch vehicle failed during a test at the Redstone Arsenal in Huntsville, Alabama. The Castor IVA booster rocket escaped from the test stand five seconds into the test. A review team was established to determine the cause of the failure and to assess the impact on the Delta flight program. (NASA Release 87-128)

August 27: NASA and the Department of Defense announced that the conceptual design phase of the National Aero-Space Plane Program was finished. The program would now move to the subsystem fabrication and test portion of the program. This phase of the engine technology development would demonstrate the ability of airbreathing engines to power a horizontal takeoff and landing vehicle. It would demonstrate that the X-30 flight vehicle was capable of sustained hypersonic flight or direct, single stage, ascent into orbit. The development of this technology would create an entirely new family of operational aerospace vehicles. (NASA Release 87-129)

• NASA Administrator Dr. James C. Fletcher issued a statement in response to Senator William Proxmire's request in Congress to abolish the Space Station program. Dr. Fletcher pointed out that the Space Station proposal has been repeatedly debated in Congress and each time has received bipartisan congressional support. He stressed that by abolishing the Space Shuttle program, the United States would abdicate its position as a world leader in space. He also emphasized that a Space Station would lead to new scientific and technological advances and would "provide infrastructure for future exploration of the solar system by manned spacecraft." (Statement by Dr. James C. Fletcher in Response to Senator Proxmire's Statement Asking for Abolishment of the Space Station, August 27, 1987)

August 30: NASA successfully test fired a redesigned shuttle booster in Utah's Wasatch Mountains. NASA officials, very pleased with the test firing, cautioned, however, that the complete success of the firing cannot be fully determined until engineers have taken the hardware apart and fully analyzed it. They added that, at any rate, this was just the first step and much work remained if the Shuttle was to make its flight date in June 1988. The 14-story test booster, built by Morton Thiokol, Inc. and known as "DM-8", was loaded with 1.1 million pounds of propellent. (*B Sun,* Aug 31/87; *NY Times,* Aug 31/87; *WSJ,* Aug 31/87; *W Post,* Aug 31/87)

During August: Data collected by the Giotto spacecraft, which flew by the Halley comet in March 1986, suggested that tiny chains of formaldehyde chemicals, contained in the comet's gassy halo, may be older than the solar system. (*LA Times,* Aug 9/87)

NASA's Marshall Space Flight Center in Huntsville, Alabama, awarded a 2.6 million contract to the city's Wyle Laboratories' Scientific Services & Systems Group to design and develop the multiple experiment processing furnace. The furnace, for processing metals and alloys in space, was to be carried aboard the Space Shuttle where astronauts would carry out their experiments. (*Def News,* Aug 10/87)

September

September 1: United States and Australia signed a 10-year agreement permitting NASA to launch scientific sounding rockets from Woomera, Australia. The first project under the new agreement would consist of five or six launches of Black Brant IX's and Nike-Black Brant V's to study Supernova 1987a during November/December 1987. (NASA Release 87-132)

September 2: NASA announced that the three Space Shuttle main engines are undergoing acceptance testing at its National Space Technology Laboratories in Mississippi. The Rocketdyne Division of Rockwell International performed the tests. The engines tested were to be used in the next Shuttle mission, STS-26, scheduled for launch in June 1988. The tests were to be completed in December 1987. (NASA Release 87-130; *CSM,* Sept 5/87)

September 4: McDonnell Douglas Astronautics Company of Houston, Texas, was awarded NASA's applications and analysis support contract for Johnson Space Center's (JSC) Mission Support Directorate, in Houston, Texas. The JSC Mission Support Directorate is responsible for space mission planning. The new contract consolidated five different existing support contracts covering a wide range of computer software development and software performance analysis. (NASA Release 87-131; *The Citizen,* Sept 6/87)

September 14: NASA agreed with most of the findings and recommendations listed in the report by the National Research Council Committee on Space Station. Agency officials disagreed, however, with the findings that the deployment of the Space Station with the current Space Transportation System was risky. NASA also found the Committee's cost estimate for back-up hardware and test program enhancements to be too high. (NASA Release 87-135; *NY Times,* Sept 15/87; *LA Times,* Sept 15/87; *H Post,* Sept 15/87)

• Officials at Goddard Space Flight Center, Greenbelt, Maryland, reported that NASA scientific satellites, the International Sun-Earth Explorers ISEE-1 and ISEE-2, were expected to reenter Earth's atmosphere two minutes apart on September 26, 1987. They were expected to create fireballs as they burned in the atmosphere over Brazil. The two orbiting spacecraft, launched on a single Delta rocket on October 22, 1977, for nearly 10 years studied fluctuations in plasma waves, the magnetic field, proton and electron density, cosmic rays, gamma ray bursts, and the solar wind in the near-Earth environment. (NASA Release 87-137)

September 15: Two firms, General Dynamics, San Diego, California, and Martin Marietta, New Orleans, Louisiana, were chosen by NASA for liquid-fueled rocket booster design studies. The liquid-fueled rocket boosters could be used on the Space Shuttle and future launch vehicles. (NASA Release 87-138; *H Chron,* Sept 17/87)

• NASA selected five crew members for STS-27, a Department of Defense Space Shuttle mission. The five crew members, scheduled for launch on Atlantis in early fall 1988, included Robert L. Gibson (Cdr., USN), Commander; Guy S. Gardner (Lt. Col., USAF), Pilot; and Mission Specialists Richard M. Mullane (Col., USAF), Jerry L. Ross (Lt. Col., USAF), and William M. Shepherd, (Cdr., USN). (NASA Release 87-139; *LA Times,* Sept 16/87; *USA Today,* Sept 16/87)

• A 15-story high Ariane 3 rocket, carrying two communications satellites, was launched into an Earth orbit from Kourou, French Guiana. The successful launch marked the reentry of Western Europe into the commercial utilization of space after a 16-month hiatus. (*LA Times,* Sept 16/87; *NY Times,* Sept 16/87)

September 17: Dr. Robert Watson, NASA Chief Scientist for the Airborne Antarctic Ozone project, reported that, midpoint of the project, both aircraft involved were performing very well and that mission scientists were very pleased with the quality of data being returned by the instruments. The Antarctic Ozone Project employed two NASA aircraft, a modified flying laboratory DC-8, and an advanced high-altitude ER-2. The goal of the project was to study the cause and nature of the ozone layer depletion. (Untitled release from NASA Headquarters, Sept 17, 1987; *NY Times,* Sept 22/87)

September 18: The Boeing Commercial Airplane Company was selected by NASA's Langley Research Center, Hampton, Virginia, for major aeronautical flight experiments. NASA tasked Boeing with providing data on the aerodynamic and operational effectiveness of a hybrid system to achieve laminar air flow control at flight conditions, particularly at high subsonic speeds. The tests were to be conducted on both commercial and military transport airplanes. (NASA Release 87-140; *Def News,* Oct 12/87)

September 25: NASA announced that researchers at the Jet Propulsion Laboratory in Pasadena, California, had successfully tested a passenger vehicle satellite antenna. The newly developed antenna, designed to be mounted on the roof of a passenger vehicle, was expected to play an important role in NASA's mobile satellite experiment program. This program and other technologies were expected to lead eventually toward a mobile satellite communications system. The first success in the program was achieved when researchers, using an antenna on moving passenger vehicle, were able to track an in-orbit satellite. (NASA Release 87-143)

• General Dynamic Corporation announced it had been awarded a contract to launch between one and three European satellites by the European Telecommunications Satellite Organization (EUTELSAT). The 26-nation European consortium selected General Dynamic's Atlas-Centaur rockets for the launching of its satellites because of their cost and reliability. The agreement called for a launch of a EUTELSAT-2 communications satellite in early 1990 from Cape Canaveral, Florida, with options for two additional launches. (*LA Times,* Sept 26/87)

September 29: NASA announced that secondary payloads for the next Space Shuttle mission would include five microgravity experiments, life sciences, atmospheric science and infrared communications experiments, as well as two student experiments. (NASA Release 87-144)

• More than 50 NASA-sponsored scientists were involved in 27 U.S./U.S.S.R. joint experiments aboard a Soviet Union Cosmos 1887 launched on this date. This mission represented the latest of the 16 agreed-upon collaborative projects under the U.S./U.S.S.R. Space Agreement signed in April 1987. (NASA Release 87-146; *O Sen Star,* Sept 29/87)

September 30: NASA Deputy Administrator Dale D. Myers announced that James T. Rose had been selected as Assistant Administrator for Commercial Programs, effective October 5, 1987. He would be responsible for NASA's commercial programs, the use of space by the U.S. private sector, the use of NASA technology by private industry, and NASA's support of innovative research by small business firms.

James T. Rose joined NASA in 1957 as a research engineer for the Vanguard Program experiments. In 1962, he became head of the Manned Spacecraft Center's Mission Planning Office for the Gemini flights. From 1964 to 1974, Rose held a series of key positions at the McDonnell Douglas Astronautics Company. He returned to NASA in 1974 as Director, Space Shuttle Engineering, in the Office of Space Flight. From 1976 to 1986, he was Director of McDonnell Douglas' Space Manufacturing Initiatives. In June 1987, he again returned to NASA to become the Director of Space in NASA's Office of Aeronautics and Space Technology. (NASA Release 87-145; *Huntsville Times,* Oct 1/87)

September 31: Martin Marietta Corporation announced that its Denver-based Space Systems group was one of the companies awarded two NASA contracts to study and design spacecraft for a mission to Mars. A contract to study the techniques of a robotic rover to traverse the Martian surface was awarded by NASA's Jet Propulsion Laboratory. NASA awarded a similar contract to the FMC Corporation of Santa Clara, California.

The second contract was awarded by NASA's Johnson Space Center (JSC), Houston, Texas. It called on Martin Marietta to study braking methods that would allow a spacecraft to orbit Mars, enter its atmosphere, and descend to the surface of the planet. The Lockheed Missiles and Space Company of Sunnyvale, California, was the other company awarded this study contract by the JSC.

Martin Marietta was the company that designed, built, and tested the twin Viking spacecraft that landed on the surface of Mars in 1976 and 1977. (*AvWk,* Oct 26/87; *B Sun,* Oct 1/87, *Rocky Mountain News,* Oct 1/87)

October

October 1: A NASA report, titled "Space Science Operations Management Concepts Study," was submitted to the House Committee on Science, Space, and Technology. The report, prepared by a 60-member study team of personnel from NASA Headquarters, field centers, other Federal agencies, and university officials, examined options for the management of Space Station science. (NASA Note to Editors, Oct 1/1987)

October 2: NASA and the U.S. Air Force signed a pricing agreement, establishing reimbursement policies for Department of Defense (DoD) flights on the Space Shuttle and NASA flights on DoD-procured expendable launch vehicles. (NASA Release 87-147)

October 3: Soviet and Japanese scientists confirmed the first positive detection of x-rays from an exploding supernova. Soviet scientists, using x-ray telescopes aboard the Soviet astrophysics module Kvant, made the detection on August 10, 1987. Japanese scientists, using smaller x-ray instruments aboard their Ginga satellite, first observed the x-rays from the supernova on August 15, 1987. (*W Post,* Oct 4/87)

October 5: NASA awarded a contract to the Computer Sciences Corporation, El Segundo, California, for operations support to the Mission Support Directorate of Johnson Space Center (JSC) in Houston, Texas. Under the contract, Computer Sciences Corporation would provide maintenance, operations, and sustaining engineering of institutional ADP facilities, networks and work stations, and user support. About 30 percent of the work would be performed by two Computer Sciences Corporation subcontractors, International Business Machines Corporation and Ford Aerospace and Communications Corporation, a unit of Ford Motor Company. (NASA Release 87-148; *WSJ,* Oct 7/87)

October 6: NASA and West Germany signed a cooperative Earth observation agreement. Under the agreement, West Germany, together with Italy, was to provide an X-band synthetic aperture radar (X-SAR) to fly with the space-borne imaging radar-c (SIR-C) on the Space Shuttle. The SIR-C will allow the Shuttle-borne imaging radar mission to view a site from different angles and at different radar frequencies. An advanced X-SAR was expected to combine later with the NASA Earth Observation System to fly on a NASA polar platform, providing a third frequency to the combined mission. (NASA Release 87-149)

October 7: NASA and the Pentagon awarded contracts to McDonnell Douglas Corporation, General Dynamics Corporation, and Rockwell International for developing the airframe of an aircraft capable of working both as a space vehicle and a hypersonic airplane. After completion of the designs by the three companies, one is to be chosen in 1990 to build two research aircraft. (*B Sun,* Oct 8/87; *H Post,* Oct 8/87)

October 8: Because of complications encountered in the nozzle fabrication process, NASA moved the test of the flight configuration of the redesigned Space Shuttle solid rocket motor (SRM) to middle or late December 1987. (NASA Release 87-151)

October 12: A Soviet space capsule landed in eastern Siberia, thousands of miles from its designated landing site in Central Asia. The capsule, a biosatellite Cosmos 1887, which carried two monkeys, rats, amphibians, fish, and insects, was launched into orbit September 29, 1987. The planned 2-week fly in space, of the capsule, was almost aborted when one of the monkeys freed his left hand on the fifth day of the mission. (*B Sun,* Oct 13/87; *NY Times,* Oct 13/87; *P Inq,* Oct 13/87)

October 19: General Dynamics Space system Division was awarded a contract by NASA and the Department of Commerce for expendable launch vehicle transportation services for meteorological spacecraft GOES (geostationary operational environmental satellites). The GOES spacecraft provided nearly continuous, high-resolution visual and infrared imaging of weather systems over large areas of the Earth. (NASA Release 87-156)

October 22: NASA announced that in the next three years it would reduce Shuttle use by launching only 19 space flights. During the same time, it would launch 49 satellites on untended rockets. About 30 non-military payloads originally planned to be carried into orbit by the Shuttle through 1995, will now be carried on untended rockets.

NASA said that it would launch five science missions in 1989, some with international cooperation. Four of these missions—the Magellan, to map Venus; the Hubble Space Telescope, in cooperation with the European Space Agency; and the ASTRO-1 ultraviolet observatory and Galileo, in cooperation with Germany to make a comprehensive survey of Jupiter and its moons—would fly on the Shuttle. A fifth mission, the Cosmic Background Explorer (COBE) to investigate cosmic background noise, was to launch on a Delta. Another scheduled Shuttle launch was the cooperative ESA/NASA Ulysses mission to observe the polar region of the Sun.

From 1990 through 1995, NASA expected to accelerate deployment of other space science missions by fully utilizing expendable launch vehicles. For example, NASA anticipated Delta launches of a Roentgen Satellite in February 1990 and the Extreme Ultraviolet Explorer in August 1991. (NASA Release 87-158; *LA Times*, Oct 23/87; *NY Times*, Oct 23/87; *W Times*, Oct 23/87)

• Martin Marietta Manned Space Systems, New Orleans, Louisiana; Rockwell International Space Transportation Systems Division, Downey, California; and United Technologies Corporation, USBI Booster Production Company, Inc., Huntsville, Alabama; were selected for contracts to perform definition studies for a proposed untended launch vehicle. The new vehicle, called Shuttle-C, would supplement the orbiter with an untended cargo element. (NASA Release 87-159)

October 26: A military satellite was lifted into orbit by a Titan 34D, the Nation's largest untended booster rocket. The launch from California's Vandenberg Air Force Base was the first successful launch of a Titan 34D rocket since February 1985. (*C Trib,* Oct 28/87; *LA Times,* Oct 27/87; *W Times,* Oct 28/87)

October 28: William V. Boynton, an Arizona University scientist charged with developing a comet penetrator-lander for NASA, announced that a prototype prober was successfully tested at the Sandia National Laboratory in Albuquerque, New Mexico. The comet penetrator, an integral part of NASA's Comet Rendezvous Asteroid Flyby mission, is an instrument-bearing missile designed to pierce a comet's surface to a depth of 3 feet or more. (NASA Release 87-161)

October 30: NASA's Office of Aeronautics and Space Technology selected five proposals for the development of space flight technology experiments as part of its Industry/University In-Space Technology program. (NASA Release 87-164)

President Ronald Reagan signed into law a $9.6 billion authorization bill that provided NASA full funding to begin work on the Space Station. (*W Post,* Oct 31/87)

November

November 2: Officials at NASA's Ames Research Center in Mountain View, California, reported that NASA's C-141 Gerard P. Kuiper Airborne Observatory would leave on October 31, 1987, for New Zealand on a month-long mission to study the supernova SN-1987a. Scientists aboard the Kuiper would make eight research flights to study the supernova, located 170,000 light years away. Supernova SN-1987a is the closest supernova to Earth since the invention of a telescope some 400 years ago. (NASA Release 87-162; *AvWk,* Nov. 9/87)

November 3: NASA Assistant Administrator for Commercial Programs James T. Rose announced new program initiatives to encourage and expand commercial space activity. Rose told some 270 executives from across the country, meeting for a three-day workshop in Nashville, Tennessee, that the new programs will build on earlier NASA commercial development policies. According to Rose, the agency already had awarded $50,000 grants to 206 small businesses for feasibility studies in this area. One of the program initiatives announced by Rose called for 28 percent space allocation for commercial secondary payloads aboard the Space Shuttle. (NASA Release 87/165; *NY Times,* Nov 5/87; *Tennessean,* Nov 4/87; *W Times,* Nov 5/87)

November 6: Appearing before the Los Angeles World Affairs Council, NASA Administrator Dr. James C. Fletcher said that building the Space Station was critical to the United States regaining initiative in space. "It is no longer a question of whether we should have long-term goals in space," Fletcher said. "The question is can we afford not to lead." (*LA Times,* Nov 7/87)

November 9: NASA announced that the allowable end-of-mission landing weight for Space Shuttle orbiters would be increased from 211,000 to 230,000 pounds. The increase means that roughly 19,000 pounds more payload can be carried aloft. The new capability was made possible by structural analysis and additional review of forces encountered by the orbiter during maneuvers shortly before landing. NASA Associate Administrator for Space Flight Rear Admiral Richard H. Truly said that this new capability would add considerable flexibility and efficiency to NASA's Space Transportation System. (NASA Release 87-168; *AvWk,* Nov 9/87; *W Post,* Nov 10/87)

• After a lengthy debate, the European Space Agency decided to fund an independent space program that would give it "full autonomy in space." Only

Great Britain, which has been pushing for cuts in the space exploration program, refused to go along. The decision was reached during the meeting in the Hague of ministers from the 13 member states, with France and Germany being the primary proponents of the new policy in space exploration.

The long-range plan adapted by the ministers included a $3.7 billion "Columbus" project calling for a crew-tended space module to be connected to the proposed U.S.-international Space Station; a free-flying human-tended Station to orbit near it; and a polar Earth-observation platform. The plan also called for about $4.4 billion for the "Hermes" miniature Space Shuttle, a reusable spacecraft with an ejectable cockpit for a crew of three, and $3.5 billion for the "Ariane" 5 rocket, which would be used to boost the Hermes into space. (*LA Times*, Nov 10/87; *NY Times*, Nov 10/87; *W Post*, Nov 13/87)

November 12: NASA announced that two Shuttle crew escape systems would be tested in November at the Naval Weapons Center in China Lake, California, using dummies and Navy volunteers. NASA is to make a decision early in 1988 on which—if either—of the two systems would be installed aboard the Space Shuttle Discovery.

One of the escape systems involved the installation of two-foot-long tractor rockets that would extract the astronauts through an open hatch. The other system consisted of a telescoping slide pole that would extend through the hatch for the crew members to slide down, away from the disabled spacecraft. The crew would be connected to the rocket or the slide pole with a lanyard.

Neither system would be of use in a Challenger-type explosion that occurred without warning. But either could function in the event of a main engine failure aborting the mission during launch. Both systems require the Shuttle to be equipped with a hatch that can be blown off the spacecraft in an emergency. (NASA Release 87-170; *H Chron*, Nov 13/87; *H Post*, Nov 13/87; *NY Times*, Nov 14/87)

November 13: The X-Wing, a NASA experimental aircraft, made its first brief venture into the air, lifting 25 feet off the runway for about 16 seconds. The X-Wing concept uses a four-bladed rotor system for vertical takeoff and landing. The first major test flight was expected to follow in about two weeks. (NASA Release 87-172; *Antelope Valley Press*, Nov 19/87)

November 15: The Pentagon released a Defense Intelligence Agency study which claimed that the Soviet Union had spent $80 billion on military space programs during the last decade. As a result, the Soviet Union was dramatically expanding its ability to boost payloads into space. The study said the Soviet Union operated about 50 types of space systems for military and civilian uses, had successfully tested a heavy-lift booster rocket in May 1987, and routinely launched five times as many rockets each year as the United States. (*P Inq*, Nov 15/87)

November 23: The National Research Council recommended, in its fifth interim report to NASA, additional testing of NASA's new solid rocket motor before resuming Shuttle launches, raising fears that the scheduled June 1988 launch could be postponed. The Council maintained, however, that the planned June 1988 launch date could still be met by concentrating on the more extensive tests and deferring some less critical ones. (*B Sun*, Nov 24/87; *NY Times*, Nov 24/87; *W Post*, Nov 24/87)

November 24: NASA released a report, requested by the U.S. Congress, on the extended-duration Space Shuttle orbiter. The report examined key aspects of a program that would allow the Space Shuttle to perform Earth-orbital missions for as long as 16 days. An extended-duration orbiter, according to the report, would be of great benefit for space science and technology. It also could prove useful in developing experiments and crew procedures in preparation for the Space Station era. (NASA Release 87-174)

November 28: The Air Force placed a military satellite, launched aboard a Titan 34D rocket, into orbit 23,000 miles above the Earth. According to a spokesman from the Federation of American Scientists, the satellite was a DSP-5R, one of a series of Defense Support Program satellites providing early warning of a hostile missile attack. (*NY Times*, Nov 30/87; *P Inq*, Nov 30/87; *W Post*, Nov 30/87; *W Times*, Nov 30/87)

During November: Martin Marietta signed an agreement with the British Ministry of Defense to launch a British Skynet 4 satellite aboard its Titan rocket in 1989. The agreement was yet another step in the U.S. effort to shift the launching of communications satellites from NASA to private companies. (*W Post*, Nov 9/87)

A General Accounting Office report found no evidence that NASA Administrator Dr. James C. Fletcher, or other space agency officials, had violated conflict-of-interest regulations in 1973, when they chose Morton Thiokol, Inc. to produce the solid-fuel booster rockets for the Space Shuttle. The investigation, carried out by the General Accounting Office, was requested last December by Senator Albert Gore Jr., Democrat of Tennessee. (*NY Times*, Nov 5/87; *W Post*, Nov 5/87)

NASA's Associate Administrator for Space Station, Andrew J. Stofan, selected Grumman Space Systems; Bethpage, New York, and Martin Marietta Astronautics Group; Denver, Colorado, as contractors for definition and preliminary designs studies of the Space Station Flight Telerobotic Servicer (FTS). The FTS is a space robot that will assist crews in the assembly, maintenance, and serving of the Space Station and visiting spacecraft. (NASA Release 87-176; *P Inq*, Nov 27/87)

A team of British and American astronomers discovered two quasars much further away from Earth than any previously known. The discovery of the two celestial objects, one that is 12 billion light years away, suggests the outer limits of the observable universe are still unknown. The new findings were largely a result of improved methods for reading photographic plates exposed by the latest telescopes. (*LA Times*, Dec 7/87; *NY Times*, Dec 3/87; *W Times*, Dec 3/87)

December

December 2: NASA announced that the long-delayed mission to Jupiter, originally scheduled for 1982, would be launched in 1989. The 5,870-pound untended Galileo spacecraft was to be launched from the Shuttle Discovery. After the Challenger accident, NASA decided to use less powerful rockets than the originally planned Centaur rockets, which are fueled with volative hydrogen. The substitution required an "audacious" rerouting of the spacecraft in order to use the gravitational force of Venus and Earth to propel Galileo toward Jupiter.

The Galileo mission was expected to reach Jupiter in 1995. Upon reaching Jupiter, the spacecraft was to release a 737-pound probe that would penetrate the atmosphere of the giant gaseous planet. In addition, the spacecraft was to spend two years orbiting Jupiter and making flybys of its four largest moons. (*CSM*, Dec 3/87; *LA Times*, Dec 3/87; *USA Today*, Dec 3/87; *W Post*, Dec 3/87)

December 10: NASA Administrator Dr. James C. Fletcher and Canadian Science Minister Frank Oberle approved the agreement on Canadian participation in the U.S. Space Station program. The terms of the preliminary agreement were spelled out in a memorandum of understanding worked out by negotiators from NASA and the Canadian Ministry of State for Science and Technology. The agreement called for cooperation between the two nations in the detailed design, development, operation, and use of the Space Station. The agreement would then be submitted for approval to the governments of the two countries. (NASA Release 87-182; *NY Times*, Dec 12/87)

December 11: The world's two largest wind tunnels became fully operational at NASA's Ames Research Center in Mountain View, California. The two tunnels, the upgraded 40-by-80 foot wind tunnel and the new, open circuit, 80-by-120 foot wind tunnel, are housed in the National Full-Scale Aerodynamics Complex. Both tunnels are used for full-scale and large-scale testing of advanced aircraft. They offer new capability for testing of rotorcraft and vertical and short-takeoff and landing aircraft (V/STOL). (NASA Release 87-180; *P Inq*, Dec 12/87)

December 12: A Soviet rocket burned like a huge fireball as it reentered the Earth's atmosphere, lighting up the skies over a broad area of the West and Midwest. The rocket was used in the launching of Soviet communications satellite Raduga 21 on December 10, 1987. (*C Trib*, Dec 14/87; *LA Times*, Dec 14/87)

December 16: NASA scientists announced that they have direct proof that exploding stars produce most of the 90-odd chemical elements heavier than hydrogen and helium. These elements are part of the composition of planets and moons in the solar system and also are found in plants and animals on Earth. The findings came from observations and analysis of the gamma-ray emissions from radioactive cobalt produced in the Supernova 1987a explosion. The observations were conducted by the Solar Maximum Mission satellite, which had been studying the Supernova since August 1987, and two balloon-borne experiments flown in October and November from Alice Springs, Australia, as part of NASA's Fall Supernova Observations Campaign. (NASA Release 87-185; *H Post,* Dec 20/87; *NY Times,* Dec 17/87; *P Inq,* Dec 20/87; *W Post,* Dec 17/87)

• A spokesman for Pentagon's Strategic Defense Initiative Organization (SDIO) stated that research and an experiment conducted by SDIO demonstrated for the first time that lasers and other high-powered electrical devices can be operated in space without heavy insulation to prevent short circuits. The experiment involved the launching of a 60-foot rocket from Wallops Island, Virginia, on December 13, 1987. Extending from the nose cone of the rocket were two probes about 39 inches apart. These probes were alternatively charged with high amounts of electrical power, up to 44,000 volts. Although there was no insulation between them, no arcing of the electrical charge between the two probes occurred, as it would commonly occur on Earth. The experiment confirmed that space is a good insulator in and of itself. (*NY Times,* Dec 17/87)

December 18: NASA officials, who met in Moscow with their Soviet counterparts December 7–13 to discuss cooperation in space, said the Soviet Union planned to launch a spacecraft to Mars in 1992. In addition, the Soviets asked the United States to modify an American spacecraft to help relay data from another Soviet mission to Mars set for launching in 1994. (*CSM,* Dec 21/87; *H Post,* Dec 20/87; *NY Times,* Dec 19/87)

December 19: An apparent technical problem forced the cancellation of the test firing of the redesigned 126-foot-long, 1.2-million-pound Space Shuttle solid rocket motor, Development Motor-9 (DM-9), to be used in the first post-Challenger flight. The failed test took place at Morton Thiokol's Wasatch Facility near Brigham City, Utah. The test was canceled one second before ignition. This was the second of at least five full-scale tests planned to check the operation of the booster's redesigned o-ring joints. The first full-duration test firing of a redesigned motor took place on August 30, 1987. (NASA Release 87-183; *P Inq,* Dec 20/87; *W Post,* Dec 20/87)

December 21: The Soviet Union launched a Soyuz TM-4 from Baikonur Space Center in Soviet Central Asia. The spacecraft, scheduled to dock with the Soviet space station Mir on December 23, 1987, carried three Soviet cosmonauts: mission commander Vladimir Titov, flight engineer Musakhi Manarov, and research scientist Anatoly Levchenko. They were expected to work together with the crew already aboard Mir for about a week, then Levchenko would return to Earth with Yuri Romanenko, who had been in the space station for 322 days. The new crew aboard Mir was expected to remain for a year, breaking Romanenko's record. (*B Sun,* Dec 22/87, Dec 29/87; *CSM,* Dec 30/87; *LA Times,* Dec 26/87; *NY Times,* Dec 22/87, Dec 27/87, Dec 29/87;   · *P Inq,* Dec 22/87, Dec 25/87, Dec 30/87; *W Post,* Dec 25/87, Dec 30/87)

December 23: NASA awarded contracts to the four aerospace firms it had selected on December 1, 1987, to design, develop, test, evaluate, and deliver the components and systems comprising the proposed crew-tended Space Station, planned for orbit in the mid-1990s. The contracts issued covered Phase I, the already approved elements of the Space Station program. Phase II would cover any potentially needed enhancements of the Space Station's capabilities. The four companies selected, each to perform a unique but interdependent portion of the total program, were Boeing Aerospace Company, Huntsville, Alabama; McDonnell Douglas Astronautics Company, with locations in Huntington Beach, California and Houston, Texas; General Electric Company, Astro-Space Division, with locations in Valley Forge, Pennsylvania and East Windsor, New Jersey; and Rocketdyne Division, Rockwell International, Canoga Park, California. (NASA Release 87-177; NASA Release 87-187; *C Trib,* Dec 2/87; *LA Times,* Dec 2/87; *NY Times,* Dec 2/87; *WSJ,* Dec 2/87; *W Post,* Dec 2/87)

• NASA announced the retirement of Philip E. Culbertson, Associate Administrator for Policy and Planning. Culbertson planned to leave NASA in mid-January to become President of the Lew Evans Foundation. (NASA Release 87-188)

• NASA announced that White Sands Space Harbor, New Mexico, was designated an alternate end-of-mission landing site for Space Shuttle missions STS-26 through STS-28. The new site was to be used for a Shuttle landing when conditions at Edwards Air Force Base in California precluded landing. (NASA Release 87- 189)

• Carver Kennedy, Vice President for Space Services at Morton Thiokol, Inc., said that engineers successfully test-fired the company's redesigned Space Shuttle booster rocket, planned for use when NASA resumed Shuttle flights in June 1988. (*WSJ,* Dec 24/87; *W Post,* Dec 24/87; *W Times,* Dec 24/87)

December 29: NASA announced that a more detailed examination of the results of the December 23 cold weather test-firing of the redesigned booster rocket revealed that an outer ring which anchors the booster's nozzle came apart during the test. The nozzle steers the Space Shuttle during the first two minutes of flight. The setback forced NASA to cancel the June 2, 1987 launch of the Space Shuttle. NASA officials said they were confident the launch would be delayed about three months, since the critical elements of the booster rocket, the redesigned field joints, functioned perfectly during the test. (*B Sun*, Dec 30/87; *NY Times*, Dec 30/87; *WSJ*, Dec 30/87; *W Post*, Dec 30/87; *W Times*, Dec 30/87)

• A Soyuz TM-3, carrying Colonel Yuri V. Romanenko and two fellow Soviet cosmonauts, Aleksander P. Aleksandrov and Anatoly Levchenko, parachuted to Earth about 50 miles from the town of Arkalyk in Soviet Kazakhstan. The three returned from the Soviet space station Mir, where Romanenko had set a new record of 326 days in orbit. (*NY Times*, Dec 30/87)

• The motor section of an MX missile containing nearly 100,000 pounds of rocket propellent exploded at a Morton Thiokol, Inc. plant killing five workers. The explosion at the company's Wasatch Operations plant, 25 miles west of Brigham City, Utah, was the fourth major accident at the plant. The cause of the explosion was not immediately known. Morton Thiokol also was the manufacturer of the Space Shuttle's booster rocket. (*NY Times*, Dec 30/87, 31/87; *P Inq*, Dec 30/87; *W Post*, Dec 30/87)

December 30: Two teams of NASA specialists began an inquiry as to why the test-firing of the reconstructed Shuttle booster rocket on December 23 failed. The same day, Congressman Manuel Lujan, Jr. of New Mexico urged NASA to find a second company, in addition to Morton Thiokol, Inc., to build the booster rocket. (*B Sun*, Dec 31/87; *H Post*, Dec 31/87; *USA Today*, Dec 31/87; *W Post*, Dec 31/87; *W Times*, Dec 31/87)

• The Soviet newspaper *Sovietskaya Rossiya* reported that two Soviet cosmonauts grew taller while in space. According to the paper, cosmonaut Yuri Romanenko grew four-tenths of an inch taller during his record ten and a half months in space, and his crewmate Alexander Alexandrov grew six-tenths of an inch. The gain in height was because of lack of gravity in space. (*NY Times*, Dec 31/87)

During December: NASA Administrator Dr. James C. Fletcher reviewed Agency activities during 1987 and declared that "NASA is on the road to recovery." He believed, "the United States can maintain a leadership role in space,.but only if the Nation stays truly committed to the space program." (NASA Release 87-184)

Industry officials said that the U.S. aerospace industry recorded the most profitable year in its history in 1987. (*P Inq*, Dec 17/87)

January

January 4: NASA released a proposal to select a prime contractor for a long-duration, human-tended scientific satellite called the Advanced X-Ray Astrophysics Facility (AXAF). The third of NASA's four proposed orbiting "great observatories", AXAF would study high-energy emissions associated with quasars, spinning neutron stars, and black holes, providing valuable information about these phenomena and serving as an important new tool for basic research in plasma physics. AXAF would also provide data on the various forms of "dark matter" in the universe, which may help determine whether the universe is an open or closed loop. AXAF could be scheduled for launch as early as 1995. The orbiting observatory would be 14 feet in diameter, 45 feet long, and would weigh 12–15 tons. It would be placed into a circular orbit 320 miles above the Earth, and would operate for about 15 years. Maintenance of AXAF would be accomplished by crews of either the Space Shuttle or the U.S. Space Station. (NASA Release 88-1)

January 4: Engineers at Morton Thiokol's Wasatch facility in Utah concluded that a design flaw had led to the failure of a previously untested Space Shuttle booster part during a full-scale test firing December 23. NASA officials delayed the flight schedule after they discovered that a portion of the booster's nozzle assembly, which guides the vehicle, had fallen apart during testing, a month earlier.

Engineers finished taking the booster motor apart and recovered all six missing pieces of the failed nozzle part. Known as the outer boot ring, the failed part anchored the booster's nozzle to a flexible rubbery "boot" that allowed the nozzle to swivel. One purpose of the boot ring was to shield the metal and rubber bearing at the core of the swivel mechanism from the intense heat of the rocket gases. During the test firing, the booster's nozzle was deliberately moved to the maximum seven degrees—some three degrees beyond what would be required during a normal Shuttle flight. NASA was looking into the possibility that the extreme steering might have caused the boot ring to come apart or delaminate.

A group of officials at the Johnson Space Center, Houston, Texas, developers of the original design test requirements for an extreme swiveling of the nozzle, studied whether the maneuver should be repeated in future tests of whatever design is adopted, or whether a less severe maneuver would be an adequate test under the more stringent, post-Challenger standards. The test failure would probably force a delay of 6 to 10 weeks in the resumption of Space Shuttle flights, delaying the STS-26 flight of Discovery to no earlier than August 1988. (NASA Release 87-190; *UPI,* Jan 4/88; *W Post,* Jan 5/88; *NY Times,* Jan 6/88)

January 11: Speaking at the American Institute of Aeronautics and Astronautics supported Aerospace Sciences Conference, Ross M. Jones, an engineer at NASA's Jet Propulsion Laboratory in Pasadena, California, presented a proposal to launch very small, low-cost scientific probes to various destinations in the solar system by using Earth-orbiting electromagnetic launchers or railguns. Electromagnetic launchers, which produce strong electromagnetic forces that accelerate projectiles to extremely high velocities, were being developed by the U.S. military for use both on the ground and in space. (NASA Release 88-4; *M News,* Jan 11/88; *W Times,* Jan 11/88; *LA Times,* Jan 11/88; *LA Star News,* Jan 12/88)

January 13: President Ronald Reagan signed a report to Congress creating a joint Department of Defense (DoD) and NASA program for the development of the Advanced Launch System (ALS). The ALS was intended to provide a launch system that would meet long-term national launch needs; would be flexible, robust, reliable, and responsive; and would drastically cut costs in all elements of the space launch system. The management plan created a joint program office headed by an Air Force program manager and a NASA deputy program manager. DoD would manage the systems engineering and integration, vehicle, logistics, and payload module. NASA would manage liquid engine systems and focused technology efforts. DoD would accept full funding responsibility for the program, with any unique civil requirements not addressed by the ALS baseline being funded by NASA. (NASA Release 88-5)

January 15: NASA announced that Dr. Raymond S. Colladay, Associate Administrator for the Office of Aeronautics and Space Technology, would leave the Agency to serve as Director of the Defense Advanced Research Projects Agency, effective February 1, 1988. (NASA Release 88-6)

January 22: EG&G Florida, Inc. was awarded a $635,529 contract to make modifications to the Space Shuttle Landing Facility (SLF). This work would include grinding a 3,500-foot section at each end of the runway to smooth the surface texture, removing cross grooves and adding longitudinal "corduroy grooving." Additionally, existing landing zone light fixtures were to be modified and markings on the entire runway and overruns repainted. These modifications were intended to enhance landing safety by reducing Space Shuttle orbiter tire wear during landing operations. The SLF is 15,000 feet long and 300 feet wide, with a 1,000-foot paved safety overrun at each end. Work was scheduled to begin January 26 and be completed by mid-March 1988. (NASA Release 88-10)

January 26: An engineer who, only hours before the Space Shuttle Challenger blew up, warned NASA that it was dangerous to launch the spacecraft in cold weather was named a winner of the Scientific Freedom and Responsibility Award. Roger M. Boisjoly, a former engineer for Morton Thiokol, Inc.—the manufacturer of the Space Shuttle solid rocket motor blamed for the Challenger accident that killed seven crew members—was honored by the American Association for the Advancement of Science (AAAS) for attempting to halt the launch and for his attempts to have the flawed rocket redesigned. The AAAS, in an announcement, cited Boisjoly "for his exemplary and repeated efforts to fulfill his professional responsibilities as an engineer by alerting others to life threatening design problems on the Challenger Space Shuttle and for steadfastly recommending against the tragic launch." Boisjoly was to receive the AAAS award during February 14 ceremonies in Boston. (*LA Star News,* Jan 27/88)

January 26: This date marked a decade of continuous operation of the International Ultraviolet Explorer (IUE). The IUE is credited with some of the most important advances in modern astronomy, including the discovery of galactic halos, monitoring volcanic activities on Io, beaming the first images ever recorded of Halley's comet from space, and monitoring the intense emissions of ultraviolet radiation from Supernova 1987A. The IUE was a joint effort of NASA, the European Space Agency (ESA), and the British Science and Engineering Research Council (SERC). Goddard scientists, engineers, and technicians designed, integrated, and tested the IUE. An ESA team built the solar array and the ground facilities near Madrid. SERC, in collaboration with the University College, London, provided four TV camera detectors for transforming the spectral displays into video signals. (NASA Release 88-9)

January 27: NASA announced August 4, 1988, as the target for launch of the next Space Shuttle mission, STS-26. This new launch target was selected following a major program assessment, subsequent to the most recent full-scale firing of the redesigned solid rocket motor (SRM) in December 1987, which revealed a design defect in the outer boot ring of the SRM case-to-nozzle joint. As a result of the assessment, NASA officials determined that the SRMs that would boost the orbiter Discovery on STS-26 would use an alternative outer boot ring configuration tested successfully during the Development Motor-8 firing in August 1987.

Rollout of Discovery to Kennedy Space Center Launch Pad 39B was scheduled for May 13, and a flight readiness firing of Discovery's main engines and liquid propulsion system was to take place on June 13 (NASA Release 88-11; *NY Times,* Jan 28/88; *W Post,* Jan 28/88; *USA Today,* Jan 28/88; *LA Times,* Jan 28/88; *P Inq,* Jan 28/88)

January 27: The *Washington Post* reported that the Martin Marietta Corporation agreed to launch communications satellites for the General Electric Corporation (G.E.) on untended Titan rockets during the next several years. According to the news account, the agreement to launch 15 satellites would establish Martin Marietta as the dominant company, developed in the two years since the January 28, 1986, explosion of the Challenger, in the domestic rocket business. The ageeement was expected to bring in revenues between $750 million and $1 billion for Martin Marietta through the early part of the next decade, one senior company official told the Post. Analysts told the newspaper that the G.E. agreement would make Martin Marietta dominant over two U.S. rivals in the commercial satellite launching business—McDonnell Douglas Corporation, manufacturer of the Delta rocket, and General Dynamics Corporation, manufacturer of the Atlas Centaur rocket. (*UPI*, Jan 27/88.)

January 27: A United Nations (U.N.) report warned of an alarming increase in the amount of space debris in low-Earth orbit since the first orbital launch in 1957. The report, prepared by Siegfried J. Bauer, Chairman of a U.N. panel on "Potentially Environmentally Detrimental Activities in Space," declared that "action on an international scale is obviously needed to deal with the global issue of space debris." Bauer, a professor at the Institute for Meteorology and Geophysics of the University of Graz, Austria, said most damage in space collisions until now had been caused by debris from "normal" activities in space. However, as a result of the first U.S. anti-satellite weapon tests, part of the Strategic Defense Initiative program, "there are already signs of 'willful' fragmentation and production of space debris," he said.

Bauer said that in the 30 years of the Space Age, since the first Soviet Sputnik of October 4, 1957, about 18,000 man-made objects had been projected into space. About 7,000 of them, larger than 8 inches—the visibility limit of radar—remain in "near-Earth space." Of these, 23 percent are satellite payloads, 10 percent are burned-out rocket stages, and 62 percent are various fragments. Only 5 percent are "active" satellites. Some 50 objects "appear to contain radioactive material." Besides the larger fragments, there are at least 2,000 objects ranging in size from 4 to 8 inches and about 50,000 in the range of 1/2 inch to 4 inches. Below that, there are "millions to billions of metal and paint chips in the millimeter and submillimeter range" which pose greater danger than would appear from their size. Bauer claimed that a half-millimeter metal chip moving at 18,500 miles per hour "could easily penetrate a space suit and even kill an astronaut." Most space junk is concentrated between 220 and 800 miles above Earth, "exactly where most of the satellites, the Space Shuttle and space stations operate."

Bauer said that at this point "the possibility of a collision with space debris is still very small, but is not completely negligible, particularly compared to the hazard from extraterrestrial material such as micrometeorites." Bauer listed several examples of damage to satellites caused by man-made space debris. (*UPI*, Jan 27/88)

January 28: NASA Administrator Dr. James C. Fletcher and Hiroyuki Osawa, President of the National Space Development Agency of Japan (NASDA), recently signed an agreement to allow NASA to directly receive data from Japan's Earth Resources Satellite (ERS-1). The agreement would allow NASA to have access to real-time data from Japan's ERS-1 synthetic aperture radar and optical sensor. NASA would receive the data at a ground station at the University of Alaska, Fairbanks. Data from the ground station would be provided to NASDA. Japan's ERS-1 was scheduled to be launched by NASDA on the H-1 launch vehicle in early 1992. (NASA Release 88-12)

• NASA commemorated the second anniversary of the Space Shuttle Challenger accident. At the Kennedy Space Center (KSC), Florida, workers preparing to resume Shuttle flights paused to pay silent tribute to the Challenger crew by halting all activity at 11:38 a.m., the time of the Challenger liftoff from launch Pad 39-B on January 28, 1986. Flags around KSC and at NASA Centers throughout the country were lowered to half staff while workers stood silent for 73 seconds, the length of the fatal Challenger flight. Just before the ceremony, KSC Director Forrest S. McCartney spoke to workers over a television circuit and loudspeakers saying: "As we make preparations to return the Space Shuttles to flight this year, it is appropriate to remember the men and women of the Challenger crew." (*UPI*, Jan 28/88; *LA Times*, Jan 29/88; *C Trib*, Jan 29/88)

January 29: It was announced in a recent NASA press briefing that there had been a total of 230 reported "close calls" since the Challenger accident. A "close call" was defined by NASA as an incident in which there is no injury, no damage, and no impact on schedule, but which possesses the potential for a more serious mishap. (NASA Release 1-29-88)

• Scientists at NASA's Goddard Space Flight Center (GSFC), Greenbelt, Maryland, successfully launched a rocket experiment aimed at simulating the effect of small comets entering the Earth's atmosphere. Following two aborted attempts, the Environmental Reactions Induced by Comets (ERIC) experiment payload was launched from GSFC's Wallops Flight Test Facility, Wallops Island, Virginia, by a two-stage suborbital Terrier-Black Brant VC rocket. At the 186-mile (300 kilometer) apogee of suborbital trajectory, the payload released a combination of water, carbon dioxide, and ice crystals into the upper atmosphere, simulating the release of gases that would result from the impact of a small comet with the Earth's atmosphere. The gas release was synchronized with the passage of NASA's Dynamic Explorer (DE) satellite over Wallops Island. The satellite photographed the atmosphere over Wallops in ultraviolet (UV) light in order to detect ionospheric depletion resulting from the release of the gases. The purpose of the experiment was to test the hypothesis that the phenomenon of ionospheric depletion—a sudden appearance of small holes in the Earth's ionosphere—is caused by the impact of small, undetected comets. The principal investigator for the project, Dr. Michael Mendillo, declared the experiment a success and noted, based on preliminary data, that the hole in the ionosphere "seemed to last longer than we had anticipated." (GSFC Release 88-5; *UPI,* Jan 29/88)

January 30: This date marked the thirtieth anniversary of the launch of Explorer 1, the first satellite put into orbit by the United States. The launch of the 31-pound satellite aboard a modified Army *Jupi*ter-C booster marked America's entry into a space race with the Soviet Union following their launch of two Sputnik satellites. In addition, Explorer 1 discovered the Van Allen radiation belts that surround the Earth. (*UPI,* Jan 30/88)

February

February 2: A DMSP F-9 meteorological observation satellite was successfully launched at 9:53 p.m. aboard an Atlas-E rocket from Vandenberg Air Force Base Space Launch Complex-3 in California. The 1,650-pound Department of Defense (DoD) satellite was placed in a near-polar orbit. (*W Times,* Feb 4/88; Aeronautics and Space Report of the President: *1988,* 171)

February 3: An experimental jetliner being developed by McDonnell Douglas Corporation debuted in a one-hour flight from the Douglas Aircraft Facility in Long Beach, California. The aircraft, with one unducted fan (UDF) engine and one ordinary jet engine, was a forerunner of the planned MD-91 and MD-92 passenger jets that McDonnell Douglas was actively marketing to airlines. The UDF engine, developed by GE, had demonstrated significant reductions in fuel consumption over conventional jet engines. McDonnell Douglas' MD-91 and MD-92 airliners were being developed to meet the expected demand among airlines for alternative, fuel-efficient aircraft during the latter 1990s. "This is as big a jump in technology as going from propellers to jets," said Walt Orlowski, program manager for the experimental aircraft. "We are at the forefront of aviation technology." (*LA Times,* Feb 3/88)

February 6: A Trident 1 missile exploded 18 seconds after a training test launch from the nuclear submarine Simón Bolívar. A brief statement issued by the Navy said the first stage fired as planned but suffered a malfunction shortly after breaching the ocean surface 50 miles off Cape Canaveral, Florida, in the Atlantic Ocean. This was the third failure for the Trident series of submarine-launched ballistic missiles since August. The Navy planned to conduct an investigation of the test failure. (*W Post,* Feb 7/88; *W Times,* Feb 8/88)

February 8: A NASA Delta 181 launch vehicle carrying a major Strategic Defense Initiative (SDI) experimental payload was launched at 5:07 p.m. EST from Cape Canaveral Air Force Station, Florida. The 6,000 pound payload, the heaviest ever carried by a Delta rocket, consisted of a main orbiting sensor platform and fifteen subsatellites released from the main spacecraft during the first 4 hours of the 12-hour experiment. The goal of the $250 million exercise was to test the ability of an orbiting sensor platform to detect and track the various components of Soviet Intercontinental Ballistic Missiles (ICBMs) in flight. Four of the subsatellites were small solid-fuel rockets whose exhaust plumes simulated the exhaust from the upper stages of Soviet ICBMs, whereas other test objects simulated real and "dummy" Soviet warhead reentry vehicles. During the mission, the main satellite and subsatellites were monitored by over 100 radar tracking stations, which simulated the ground-based component of an antimissile defense system. In addition, the exercise also included the detection, by sensors on the satellite, of a 40-foot sounding rocket launched from Kauai, Hawaii. The experiment was declared a success despite the failure of an infrared sensor. The Delta launch, originally scheduled for February 4, was delayed five days because of an apparent valve malfunction on the Delta. Data gathered by the satellite during the 12-hour exercise would be transmitted to Earth over a 10-day period. (SSR 1988 008A-D; *UPI*, Feb 9/88; *NY Times*, Feb 9/88; *W Post*, Feb 9/88; *W Times*, Feb 9/88; *USA Today*, Feb 9/88)

• NASA announced that Andrew J. Stofan, Associate Administrator for the Space Station, would retire from the Agency on April 1. Stofan's career with NASA spanned nearly 30 years. He had directed the Space Station program since June 1986. During his tenure, the program underwent a major cost review which culminated in an Administration-approved plan to develop the Space Station in two phases. Stofan also oversaw the establishment of the Space Station program office in Reston, Virginia.(NASA Release 88-16)

February 11: The White House officially unveiled a comprehensive "Space Policy and Commercial Space Initiative to Begin the Next Century," intended to assure U.S. space leadership into the 21st century. The release of the space policy had been delayed for several days because of a dispute among NASA, the President's Economic Policy Council, and Congress over aspects of the commercial space initiative. The most controversial component of the initiative was the proposed Industrial Space Facility (ISF), an orbiting, Shuttle-tended materials processing laboratory that would be jointly used by NASA and by private companies. The ISF was being promoted by some members of Congress and the Economic Policy Council as an interim step toward the permanently crew-tended Space Station. NASA Administrator Dr. James C. Fletcher submitted a "minority report" to the president opposing the policy Council's recommendation that NASA immediately commit to leasing the ISF from Space Industries, Inc., a commercial space venture. In response to the Administrator's concerns, the commercial space policy statement recommended competitive bidding to build the ISF and a reconfiguration of the facility as a crew-tended laboratory.

A draft of the 39-page commercial space initiative included a report by the National Security Council on national security implications of commercializing space, and legislative and administrative proposals by President Ronald Reagan's Economic Policy Council on how to promote commercialization of space. Among other key proposals were the promotion of a private launch vehicle industry and the establishment of limits on the amount of third-party liability insurance coverage commercial space projects would have to carry.

President Reagan also announced a program to develop the technologies needed to conduct human-assisted exploration of the solar system, including a mission to Mars. The program, labeled Project Pathfinder, would allot NASA $100 million in fiscal year 1989 to begin developing technologies for tended missions to the Moon and Mars. (*Presidential Directive on National Space Policy: Fact Sheet,* Feb 11/88; *WH Release,* Feb 11/88; *NY Times,* Jan 24/88; *W Times,* Jan 27/88; *AvWk,* Feb 1/88)

February 15: NASA successfully launched the first of three suborbital rocket experiments to study Supernova 1987A. The rocket, launched from the Woomera, Australia Range, carried an x-ray spectrometer that observed x-ray emissions in the 0.3-7 keV band. Two additional rocket experiments were scheduled for launch during a window lasting until March 20. An earlier series of sounding rocket observations of Supernova 1987A were conducted from Woomera during November and December 1987. Both campaigns used Black Brant IX two-stage, solid-fuel rockets. (LRC Release 88-7)

February 24: A segment of a Space Shuttle solid rocket motor was test fired at Morton Thiokol's Wasatch facility in Utah in a test of the solid rocket motor nozzle joint. The overall test objective was to evaluate performance of the redesigned solid rocket motor case-to-nozzle joint that had an insulation adhesive defect and damaged o-ring in order to determine the fail-safe performance of the redesigned joint. The redesigned case-to-nozzle joint included 100 added radial bolts, adhesively bonded insulation surfaces, and an added "wiper" o-ring designed to keep the adhesive on the insulation surfaces during assembly. The test was part of the Shuttle motor redesign program. (NASA Release 88-23)

• NASA announced that it would lease to the U.S. Government unassigned space aboard a commercially developed space facility. The leasing agreement would apply for five years beginning at the end of fiscal year 1993. At least 30 percent of the facility would be available for commercial use.

The facility would be deployed, checked out, and serviced periodically on orbit by Space Shuttle crews. It would provide a crew-tended, shirtsleeve workspace and would also be able to operate in an untended, free-flying mode, providing a microgravity environment for periods of four to six months. NASA would initiate an open competition to be managed by the Marshall Space Flight Center, Huntsville, Alabama. (NASA Release 88-25)

February 26: NASA announced the appointment of Manuel (Manny) Peralta as Associate Administrator for Management, effective February 28. He had been acting in this post since April 28, 1987, replacing June Gibbs Brown, nominated by the President to become Inspector General for the Department of Defense. Prior to joining NASA, Peralta had over 30 years of industry experience in business, engineering, and project management, most recently as a senior executive with Exxon, responsible for implementation of world-wide capital projects and the European Engineering Office. (NASA Release 87-26 and Key Personnel Change Announcement, Feb 25/88)

February 29: NASA announced plans to acquire a Boeing 747-100 jetliner to serve as a second Space Shuttle Carrier Aircraft (SCA) for the space transportation system. A letter contract was signed with Boeing Military Airplane Company, a division of the Boeing Company, Seattle, to reserve the aircraft for NASA use. The additional SCA would provide increased ferrying capability and eliminate a potential single-point failure in the space transportation system.

The 231-foot long aircraft would be modified to carry Shuttle orbiter vehicles from landing sites to orbiter processing facilities at the Kennedy Space Center, Florida. The 747-100, which is nearly identical to the original SCA, was selected to minimize costs associated with modifications and operation. (NASA Release 88-28; JSC Release 88-005)

During February: Aviation Week Space Technology reported that the Soviet Union was assessing a major expansion of its untended Mars balloon/rover mission planned for 1994. Soviet officials were evaluating the possibility of significantly increasing the payload capability of its dual Mars orbiter/lander spacecraft by using aerobraking instead of rocket propulsion to enter Martian orbit. Aerobraking would enable the Soviets to dispense with rockets needed to decelerate the spacecraft sufficiently to enter Martian orbit, thereby making available an additional 3,300 pounds of payload capacity.

An approved aerobraking maneuver would give the Soviets several new features to the balloon/rover mission, including: a 110-pound subsatellite that would be released into Martian orbit to provide Mars gravity data; 10 "mete-orbeacons" dropped on the surface that would return data on temperature, pressure, and wind velocity; two penetrators that would dive as deep as 16 feet into Mars' surface to return data on chemical composition, soil temperature, and water vapor content; a high resolution orbiter imaging system; and two Earth-return vehicles, which would serve as test vehicles for a future automated soil sample return mission. (*AvWk,* Feb 29/88)

• Using NASA's new supercomputer at the Numerical Aerodynamics Simulation Facility, Dr. Kozo Fujii, a research fellow at NASA's Ames Research Center, Mountain View, California, developed a *computer model* that simulates the air-flow field physics associated with vortex breakdown and provides new insights into its causes. Vortex breakdown is a complex phenomenon that can cause loss of lift and control for high-performance aircraft.

Understanding and eventually controlling vortex breakdown would improve maneuverability and safety for high-performance aircraft. (NASA Release 88-13)

March

March 1: NASA announced that an international team of researchers, coordinated by NASA's Jet Propulsion Laboratory, Pasadena, California, was using satellites to study how the Earth's tectonic plates move in the Caribbean Sea and Central and South America. The experiment, spanning several years, would allow researchers to chart the motions of dozens of land sites in 16 countries using signals from orbiting navigation satellites. Research teams carreied electronic receivers designed to pick up signals from seven NAVS-TAR satellites operated by the U.S. Department of Defense, as part of the Global Positioning System (GPS) program. By locating receiver equipment at established survey markers in South America and the Caribbean, and collecting GPS data from up to four GPS satellites passing overhead, researchers calculated the location of receiving sites on the ground and are planned to compare measurements over time to determine tectonic displacement.

The plates studied included the Caribbean plate, in portions of the southern Caribbean Sea; the Cocos plate, in the Pacific Ocean off the west coast of Central America; the Nazca plate, off the west coast of South America; and the South American plate, upon which lies most of the South American continent. The research was funded by NASA's Office of Space Science and Applications, Earth Science and Applications Division. (NASA Release 88-30)

March 3: NASA Administrator Dr. James C. Fletcher appeared before the House Subcommittee on Space Science and Applications on the agency"s proposed 1989 budget. He stressed that 1989 would be a crucial year for NASA, a year that would "make or break" the Nation's space program. The proposed budget included funds to increase the rate of Shuttle flights in 1989 in order to "fly off" the backlog of vital defense and science missions. A build up of funding for Space Station hardware development was also requested. Dr. Fletcher testified that the requested additional funds for the Space Station were the minimum necessary to avoid the disbandment of development teams and the indefinite deferral or cancellation of that program. He also warned that if funds for advanced technology were not approved, the necessary technological foundation for future achievements would not be built, and the goal of long-term U.S. space leadership would "become an idle dream." (1989 NASA Authorization Hearings Before the House Subcommittee on Space Science and Applications, Mar 3/88; NASA Release 88-31; *AP,* Mar 22/88; *W Times,* Mar 23/88; LA Times, Mar 23/88)

• NASA Administrator Dr. James C. Fletcher announced that James B. Odom was the new Associate Administrator for Space Station, NASA Headquarters, Washington, D.C. He replaced Andrew J. Stofan, who recently announced he was leaving the Agency on April 1. Odom joined the U.S. Army's rocket research and development team at Redstone Arsenal, Alabama, in 1956, as a systems engineer. He transferred to the Marshall Space Flight Center in 1959, where he held various engineering and technical management positions, including that of Chief of the Engineering and Test Operations Branch for the second stage of the Saturn V launch 3 vehicle. He was appointed Manager of the External Tank Project in the Space Shuttle Projects Office in 1972 and became Deputy Manager for Production and Logistics in the Shuttle Projects Office in 1982. He was appointed Manager of the Space Telescope Office in 1983. Odom is the recipient of numerous service awards. (NASA Release 88-32)

March 7: In response to a civil suit brought under the Freedom of Information Act by the Associated Press and six other news organizations, the Justice Department disclosed documentation showing that the Federal Government and Morton Thiokol paid $7,735,000 in cash and annuities to settle all claims with the families of four of the crew members who died in the explosion of the Shuttle Challenger. Thiokol, maker of the faulty booster rocket blamed for the January 28, 1986, explosion, paid $4,641,000. The Government's share of the settlement was $3,094,000. The settlements were reached on December 29, 1986, with the immediate survivors of the spacecraft crew members, Francis R. Scobee, Ellison Onizuka, Gregory B. Jarvis, and Sharon Christa McAuliffe. Settlements had been reached among Thiokol and two other families—those of Judith Resnik and Ronald E. McNair—with no Government contribution. A claim by the widow of Michael J. Smith against Thiokol was still pending in Federal court. The Federal Government maintained that it cannot be held liable for the deaths of civilian and military government employees who die on duty. (*UPI,* Mar 8/88; *NY Times,* Mar 8/88; *W Post,* Mar 8/88; *WSJ,* Mar 8/88; *USA Today,* Mar 8/88; *LA Times,* Mar 8/88; *W Times,* Mar 8/88; *C Trib,* Mar 8/88; *P Inq,* Mar 8/88; *AP,* Mar 13/88)

March 8: NASA announced that preparations were under way to bring on line two new Space Shuttle abort-landing sites in northwestern Africa. These sites would be used as contingency landing facilities in the event of a transatlantic abort during the launch of STS-26 and subsequent missions. The new sites would be located at Ben Guerir (40 miles north of Marrakech), Morocco, and Banjul, The Gambia.

With the addition of these new sites, and the continued use of sites at Zaragoza Air Base and Moron Air Base in Spain, the total number of transatlantic abort sites came to four. The Ben Guerir, Banjul, and Moron sites would be staffed with 30 to 40 NASA and contractor personnel during the launch of

STS-26. Robert Fleming, program manager for the contingency landing site program at Kennedy Space Center, said, "These sites must be operational and all requirements met in time for the next mission. We hope we never have to use them. But they will be ready." (NASA Release 88-34)

March 10: NASA announced it was designing a sophisticated radar receiving system to help image and track Arctic ice flows and study the remote areas of Alaska and its surrounding seas. The overall system, including a 10-meter receiving antenna called the Alaska SAR Facility, was to be located at the University of Alaska, Fairbanks (UAF), managed by the university Geophysical Institute, and implemented by NASA's Jet Propulsion Laboratory (JPL). The system initially was to receive signals from three satellites carrying synthetic aperture radar (SAR), which penetrates thick cloud cover and produces data for high resolution images.

The facility was scheduled to be fully operational in time for the April 1990 launch of the ESA satellite E-ERS 1. The facility was to receive data from Japan's J-ERS 1 satellite and the Canadian-led multinational Radarsat satellite. NASA had assembled an ad hoc team of investigators, cochaired by project scientists from JPL and the University of Alaska, to assist in facility development. (NASA Release 88-35)

• NASA announced that George W.S. Abbey was to be the new Deputy Associate Administrator for the Office of Space Flight, NASA Headquarters, effective immediately. Abbey had been Special Assistant to the Director, Johnson Space Center, Houston, Texas, since November 1987. He began his NASA career in 1967 as Technical Assistant to the Manager of the Apollo Spacecraft Program. He became technical assistant to the Center Director in 1969, Director of Flight Operations in 1976, and Director of the Flight Crew Operations Directorate in 1983. He is a recipient of numerous honors and service awards. (NASA Release 88-36)

March 11: A European Space Agency (ESA) Ariane 3 launch vehicle was successfully launched from Kourou Launch Center in French Guiana carrying two communications satellite payloads, including a U.S. satellite. The launching had been scheduled for December but was delayed for extra tests after abnormalities were detected in the oxygen and liquid hydrogen third-stage engine. Twenty minutes after launch, the Ariane 3 released the French Telecom 1C and GTE's Spacenet III Geostar R01 communications satellites for auto-boost into geostationary orbit, 22,554 miles above the Equator. The spacenet satellite was expected to begin its boost toward geostationary orbit at 87 degrees west longitude on Wednesday. (FBIS-WEU-88-049, Mar 14/88; SSR 1988 018A-D; *NY Times,* Mar 13/88; *W Post,* Mar 12/88; *LA Times,* Mar 12/88; *C Trib,* Mar 13/88)

March 12: NASA's C-141 Gerard P. Kuiper Airborne Observatory (KAO) departed NASA's Ames Research Center, Mountain View, California, for Guam to study the Sun's atmosphere during the total solar eclipse of March 17–18. A team of seven scientists aboard KAO was to make a flight through the shadow of the Moon as it moves across the Pacific Ocean. The scientists were interested in the few seconds at the beginning and end of the solar eclipse when the Sun would be almost completely covered by the Moon and when the edge of the solar disk would begin to reappear from behind the Moon. The Kuiper would intercept the path of the eclipse about 600 miles northwest of Guam. The Kuiper, a modified Lockheed C-141 jet transport aircraft, is fitted with a 36-inch diameter telescope and flies at 41,000 to 45,000 feet. Measurements of the solar atmosphere would be made at the far-infrared wavelengths of 30, 50, 100, 200, 400, and 800 micrometers, representing the first attempt to simultaneously record these wavelengths from the Sun on such a narrowly defined spatial scale. (NASA Release 88-37; ARC Release 88-08)

March 15: A report released by an international panel of over 100 scientists assembled by NASA showed that the depletion of the Earth's ozone layer over the Northern Hemisphere had increased at a much faster rate than previously believed. The report, prepared over 17 months, provided definitive evidence showing that the ozone layer, which shields life on Earth from harmful ultraviolet radiation, had been depleted by an average of about 2.3 percent, since 1969 over most of the United States, and by as much as 5 percent over the South Pole. The observed ozone loss was at least twice as large as scientists had predicted and was attributed primarily to the continued release of chlorofluorocarbons (CFCs).

The NASA study group reexamined years of ground-based and satellite observations of the atmosphere in order to remove the normal biennial fluctuations of many atmospheric characteristics and the effects of the 11-year solar cycle. The ground-based observations were limited to mid-latitudes in the Northern Hemisphere because there were few stations elsewhere in the world. This was the first effort by scientists to separate the effects of solar and biennial fluctuations from the monthly data at different latitudes. The release of the study results followed ratification by the U.S. Congress of the Montreal Protocol, an agreement endorsed by 46 countries, which calls for a 50 percent reduction in emissions of CFCs by the end of the century. (*UPI,* Mar 15/88; *NY Times,* Mar 16/88; *W Post,* Mar 16/88; *W Times,* Mar 16/88; *USA Today,* Mar 16/88; *WSJ,* Mar 16/88; *CSM,* Mar 16/88; *B Sun,* Mar 16/88; *P Inq,* Mar 16/88; *C Trib,* Mar 16/88; *LA Times,* Mar 21/88)

March 15: NASA issued an updated fleet manifest reflecting current planning for primary payloads for Space Shuttle missions and expendable launch vehicles through fiscal year 1993. Two interplanetary missions were planned for launch in 1989: Magellan, a mission to map Venus; and Galileo, a cooperative project with Germany to survey *Jupi*ter and its moons. The Hubble Space Telescope also maintained the flight assignment date of June 1989. Astro-1, a Spacelab mission designed to study the universe in the ultraviolet spectrum, was being reconfigured to enhance the study of Supernova 1987a and was slated to fly on STS-35 in November 1989. Two additional Spacelab missions were assigned flights for September 1990—a Spacelab Life Sciences mission in March and the first Atmospheric Laboratory for Applications and Science (ATLAS-1) mission. The Gamma Ray Observatory was moved forward in the projected schedule for a March 1990 launch, and the Ulysses mission to study the Sun and its environment remained at the projected October 1990 launch date.

Another important addition to the manifest was a mission to retrieve the Long Duration Exposure Facility (LDEF) in July 1989. The manifest was also designed to support the commercial space initiative included in the National Space Policy. In addition to the Shuttle launches, 35 launch vehicle launches were planned through fiscal year 1993. (*Payload Flight Assignments—NASA Mixed Fleet*, Mar/88; NASA Release 88-38)

March 16: Congress ordered a temporary halt to NASA's efforts to lease a commercial crew-tended space laboratory, citing several irregularities in NASA's handling of the procurement process and questioning the space agency's abrupt reversal of its prior opposition to the facility. The chairmen of the House appropriations subcommittee for NASA, Representative Edward Boland, and the Chairman of the House authorization subcommittee for NASA, Representative Bill Nelson, agreed to freeze funding for the facility until these issues could be addressed. Several ranking members of the House and Senate had written NASA to complain that the Agency was moving ahead with a contract without congressional authorization. (*H Post*, Mar 24/88; *AvWk*, Mar 28/88)

• Morton Thiokol, Inc. was fined by Utah state officials for six safety violations in connection with a December 29, 1987, fire that killed five workers and destroyed a missile casting building at Thiokol's Wasatch facility near Salt Lake City. The Utah Occupational and Safety Health Division sought fines of $31,700 against Thiokol, citing violations of safety rules related to protection of workers, deviations from approved procedures, and failure to take precautions against static electricity buildup in a rocket-casting facility. An Air Force investigation into the immediate cause of the fire, which occurred during the casting of solid rocket propellant into the metal casing of an MX missile segment, was pending. (*UPI*, Mar 16/88; *W Post*, Mar 17/88; *NY Times*, Mar 17/88; *WSJ*, Mar 17/88; *W Times*, Mar 17/88)

March 16: NASA began flight tests of a new design of low-noise, energy efficient aircraft propellers. The tests, undertaken jointly by NASA and Lockheed Corporation as part of the Propfan Test Assessment (PTA) program, were to determine whether large, unducted propellers with a radically swept design are a feasible alternative to higher-cost turbofan (conventional jet) propulsion systems.

Researchers estimated that a thoroughly reworked version of the old propeller aircraft would save airlines 15 to 30 percent a year in fuel costs, compared with the most advanced turbofan-powered aircraft flying in the 1990s, and still fly as fast as conventional airliners. (NASA Release 88-40)

• The Soviet Space Research Institute and the Soviet government agreed to place aboard a spacecraft, destined for the Martian moon Phobos, a plaque commemorating discovery of the Moon and its sister moon Deimos in 1877 by Asaph Hall, an American astronomer at the U.S. Naval Observatory. The Soviet Phobos mission, due for launch in July 1988, involved two Soviet spacecraft. They would reach the vicinity of Mars in early 1989 and be placed in elliptical orbits around the planet. Both would approach Phobos in sequence and release landing probes. (NASA Release 88-41)

March 18: NASA concluded negotiations with the ESA on a bilateral memorandum of understanding (MOU) for cooperation in the design, development, operation, and use of the permanently crew-tended civil international Space Station complex. Under the terms of the new MOU, ESA would provide the Columbus laboratory module to the international Space Station complex. This permanently attached, pressurized module would support approximately 40 single equipped racks for payloads and storage. ESA would also provide an untended, free-flying polar platform for Earth observation experiments, complementing the polar platform under development by the United States. As a third element, ESA would provide a man-tended free flyer (MTFF) designed to accommodate long-duration microgravity applications in the fields of fluid physics, life, and material sciences. (NASA Release 88-44)

March 22: NASA unveiled plans for an ambitious program to study the Earth as a single ecological system, the *Los Angeles Times* reported. Called the Earth Observing System, the U.S. portion of the multinational project would cost well above $2 billion and would use a wide range of orbiting laboratories, including free flying platforms and facilities aboard the international space station. The heart of the program would include four platforms—two from the United States, one from Europe, and one from Japan—that would be launched into polar orbit around the Earth by the participating countries. The program, which was still in the planning stage, was described by NASA science division spokesman Charles Redmond as "a critical element in NASA's strategic plan." (*LA Times,* Mar 23/88)

March 25: The San Marco D/L spacecraft was successfully launched on a U.S. Scout launch vehicle from the San Marco Equatorial Range in Kenya.

The launch, originally planned for March 18, had been delayed to permit replacement of a yaw rate gyro in the Scout. Five scientific instruments—three from the United States, one from the Federal Republic of Germany, and one from Italy—were carried on the spacecraft to study Earth's lower atmosphere. (SSR 1988 026B; NASA DAR, Mar 3/88; NASA Release Mar 2/88; LRC Release 88-8, 88-8A)

March 30: NASA and the Council of Chief State School Officers (CCSSO) invited America's students to participate in a national competition to name NASA''s replacement Space Shuttle orbiter. NASA specified that the name chosen not only should identify an American spacecraft but also should capture the spirit of America's mission in space. In honor of the seven crew members lost in the Challenger accident, the name "Challenger" has been retired.

House Joint Resolution 559, introduced March 10, 1986, by Congressman Tom Lewis (R-Florida), called for the name of the replacement orbiter to be selected from suggestions submitted by students. On October 30, 1987, Congress authorized the NASA Administrator to select a name for the new orbiter "from among suggestions submitted by students in elementary and secondary schools." The new orbiter, designated OV-105, under construction by Rockwell International in California, was scheduled for completion in April 1991. (NASA Release 88-46).

During March: The Air Force and NASA debated possible new destruct guidelines covering when a malfunctioning Shuttle and crew would have to be blown up to prevent greater loss of life on the ground. In the wake of the Challenger accident, the debate over new Shuttle range safety guidelines focused on increasing public safety while at the same time giving the crew of a malfunctioning Shuttle as long as possible to overcome the problem and "fly it out" before a destruct order would have to be issued. As a result, the Air Force suggested barring all non-essential personnel—VIPs, journalists and spectators—from the Kennedy Space Center for future Shuttle launches. "It's the kind of thing [where] you wish you could hide your head in the sand and pretend like it didn't have to exist, that there was never going to be any problem," said veteran Robert "Hoot" Gibson, Commander of the second post-Challenger Shuttle flight. (*UPI,* Mar 5/88, *H Chron,* Mar 6/88)

April

April 2: NASA announced that it would remove and inspect the liquid oxygen turbopumps aboard the Shuttle Discovery, a move that was expected to delay Discovery's launch beyond the scheduled date of August 4. The pumps will be inspected to ensure that critical bolts in the devices are properly tightened and have not been stripped. Delays resulting from the inspection are expected to push back Discovery's launch date until late August. (*UPI,* Apr 2/88; *W Post,* Apr 3/88; *USA Today,* Apr 4/88; *P Inq,* Apr 4/88)

April 5: NASA and the ESA announced the selection of scientific investigations to be conducted on two spacecraft as part of a joint NASA-ESA solar terrestrial research program. More than 130 U.S. scientists were scheduled to participate in the Solar and Heliospheric Observatory (SOHO) and Cluster missions. Two spacecraft systems were to be built by ESA with instruments supplied by both ESA and NASA. The Cluster mission, involving a set of four satellites, was planned for launch on an ESA Ariane launch vehicle in December 1995. The Cluster would investigate the space plasma three-dimensional microphysics phenomena in Earth's magnetic field environment, using four identical spacecraft in polar orbits. SOHO, planned for launch by NASA on the Shuttle or on an expendable launch vehicle in March 1995, would be placed at the Earth-Sun La Grangian point and investigate the physical processes that form and heat the solar corona and give rise to the solar wind.

The SOHO and Cluster missions are an outgrowth of the International Solar Terrestrial Physics (ISTP) program, a major international program in solar and space physics organized in 1985 by NASA, the Japanese Institute for Space and Astronautical Science (ISAS), and ESA. (NASA Release 88-48)

April 7: NASA selected a telescoping pole as the egress method for the Space Shuttle's new crew escape system. The escape system, to be carried on all future Shuttle flights, would provide crew escape capability from the orbiter during controlled, gliding flight following failures or difficulties during ascent or entry where landing at a suitable landing field could not be achieved. Tests conducted in February and March, using a fixed pole extending through a hatch-like opening in a C-141 aircraft, demonstrated that the pole would provide adequate orbiter clearance in an emergency egress situation. The telescoping pole system had been chosen over an alternative tractor rocket extraction system. (NASA Release 88-50; JSC Release 88-014; *AP,* Apr 8/88; *NY Times,* Apr 8/88; *W Post,* Apr 8/88; *USA Today,* Apr 8/88; *LA Times, C Trib,* Apr 9/88; Apr 8/88; *P Inq,* Apr 8/88; *H Post,* Apr 8/88; *H Chron,* Apr 8/88)

April 11: Astronomers at the University of Hawaii at Manoa announced the discovery of the oldest and most distant galaxy ever seen. The galaxy, called 0902+34, was estimated to be 10 times the size of the Milky Way and 12 billion light years from Earth. Measurements indicated that the galaxy was formed much earlier in the history of the universe than had been generally thought possible, requiring a revision of accepted theories on the distribution of matter in the early universe and the formation of galaxies. The galaxy was discovered by Dr. Simon J. Lilly, a staff astronomer at the University of Hawaii at Manoa. (*NY Times,* Apr 12/88; *W Post,* Apr 12/88; *W Times,* Apr 12/88; *P Inq,* Apr 12/88)

April 14: NASA Administrator Dr. James C. Fletcher stated that the Moon, rather than Mars, may be the best initial destination for possible U.S./U.S.S.R. human-assisted missions. "Going to the Moon together would give the two leading spacefaring nations in the world an opportunity to build a stable base for further cooperation, which could, one day, lead to a cooperative mission to Mars," he said. Speaking before the National Space Symposium in Colorado Springs, Colorado, Fletcher cited timing, cooperative experience, and technical readiness as three crucial factors favoring the Moon for an initial cooperative crew-tended mission.

Dr. Fletcher described 1988 as "perhaps the most critical year in the history of the U.S. civil space program," and he refuted the notion that American space leadership was lost.

Dr. Fletcher said the Administration's fiscal year 1989 budget request for NASA provided the resources to reestablish U.S. leadership in space. (Excerpts from Remarks Prepared for Delivery, National Space Symposium, U.S. Space Foundation, Colorado Springs, Colorado, Apr 14/88; NASA Release 88-52)

April 18: NASA submitted to the Congress its acquisition plan for the Space Shuttle Advanced Solid Rocket Motor (ASRM). The plan contained NASA's strategy for a full and open competition to introduce an ASRM into the Shuttle system. NASA contended that a new booster design would result in substantive improvements in flight safety, reliability, and performance over the current SRMs produced by contractor Morton Thiokol, Inc.

The ASRM design considered by NASA would be an advanced version of the segmented boosters currently in use in the Shuttle program. The new boosters would have a more reliable field joint design and would be more powerful than the current SRMs, providing for a 12,000-pound increase in the Shuttle's payload capacity. Additionally, the ASRMs would preclude the necessity for throttling the Space Shuttle's main engines during the period of maximum dynamic pressure, thereby reducing or eliminating about 175 "criticality 1/1r" failure modes for the Shuttle system.

An important aspect of NASA's ASRM acquisition plan was the proposal that the boosters be developed in a new Government-owned contractor operated (GOCO) facility and that the contract be subjected to competitive bidding. In response to protests by Morton Thiokol and the Utah congressional delegation, led by Senator Jake Garn, NASA altered the proposal to include the possibility of building the ASRM in a currently existing private facility. The overall cost to develop the ASRM was estimated at just under $1 billion, including modern tooling and equipment, plus $200–$300 million for construction of facilities. Under the current production schedule, the first ASRMs would be flight certified by 1993. Thiokol announced it would bid aggressively to build the ASRM at its existing facilities in Utah. (NASA Release 88-54; *UPI,* Apr 18/88; *AP,* Apr 19/88; *USA Today,* Apr 19/88; *W Times,* Apr 19/88; *NY Times,* Apr 20/88; *W Post,* Apr 20/88; *WSJ,* Apr 20/88;)

April 20: The third full-duration test firing of NASA's redesigned Space Shuttle solid rocket motor was carried out at Morton Thiokol's Wasatch Facility near Brigham City, Utah. The 126-foot long, 1.2 million-pound motor, designated Qualification Motor-6 (QM-6), underwent a full-duration horizontal test firing of two minutes. The test, tentatively labeled a success, was the third of five test firings to be carried out prior to the scheduled resumption of Shuttle flights in August 1988. (NASA Release 88-51; MSFC Releases 88-43, 88-44; *AP,* Apr 20/88; *UPI,* Apr 20/88; *NY Times,* Apr 21/88; *W Post,* Apr 21/88; *USA Today,* Apr 21/88; *W Times,* Apr 21/88)

April 21: The Canadian government announced that it would become an active partner in the international Space Station Program and would provide equipment worth $1.2 billion to the project. Canada's main equipment contribution to the Space Station would be a roving space robot that would serve as the "arms and legs" of the Station. The robot, which would be comparable to but more advanced than the "Canadarm" currently carried on U.S. Space Shuttles, would be a critical tool in Station construction and docking activities. In return for Canada's contribution, Canadian astronauts would be entitled to six-month tours of duty on the Station every two years. Canadian space officials announced plans to expand their astronaut corps in anticipation of the Canadian presence on the Space Station. (*Tor Star,* Apr 22/88)

April 25: Two U.S. Navy navigation satellites were launched from Vandenberg Air Force Base, California, into a 600-mile circular polar orbit aboard a NASA Scout rocket. The two Oscar satellites were part of the U.S. Navy's long-established, continuous all-weather global navigation system. Made available to non-Navy users in 1967, the spacecraft had since been adapted for diverse civilian uses such as commercial shipping, charting of offshore oil and mineral deposits, and land survey projects. The system provided position information within one-tenth of a nautical mile anywhere in the world. This marked Scout's 100th flight and followed closely on the heels of the Scout San Marco D/L launch, which took place on March 25 from the San Marco Range platform in the Indian Ocean. (SSR 1988 033A-G; NASA Release 88-55)

April 28: The fifth and final short-duration solid rocket motor Nozzle Joint Environment Simulator (NJES) firing test was carried out. The overall objective of the test, part of the Space Shuttle solid rocket motor redesign program, was to evaluate the fail-safe performance of the redesigned case-to-nozzle joint with verified defects. (NASA Release 88-56)

May

May 4: A series of powerful explosions destroyed the Pacific Engineering & Production Company oxidizer plant near Las Vegas, Nevada, killing two plant employees and injuring over 350 people. The plant was a major supplier of the ammonium perchlorate oxidizer used aboard the Space Shuttle's solid rocket motors and other military and civilian boosters. The explosions produced flames that shot 100 feet into the air and created a 400 foot-wide crater at the facility site. Seismographs at NASA's Jet Propulsion Laboratory, 200 miles away in Pasadena, California, detected two of the blasts registered at 3.0 and 3.5 on the Richter scale. The cause of the explosions was investigated.

The destruction of the oxidizer plant raised concerns among NASA and contractor Morton Thiokol over the availability of solid fuel oxidizer beyond the next four Space Shuttle flights. A prolonged interruption of Pacific Engineering's production activities would also threaten the military's missile production schedule. (*UPI,* May 5/88; *AP,* May 5/88; *NY Times,* May 5/88; *W Post,* May 5/88; *WSJ,* May 5/88; *USA Today,* May 5/88; *W Times,* May 5/88; *B Sun,* May 5/88)

May 5: NASA Administrator Dr. James C. Fletcher voiced alarm over the House of Representative's resolution for funding of NASA's fiscal year 1989 budget. Speaking in Washington before a symposium on "Science Education: A Challenge for Excellence in America's Future," Dr. Fletcher issued his toughest warning yet about the consequences of reducing the NASA funding levels proposed by the Administration. the Administrator warned that "the civil space program will be stopped in its tracks" if the House funding figure of $10.2 billion—cut by over $1 billion from the Administration proposal—were to be passed. More specifically, Dr. Fletcher predicted the reduced budget would kill the Space Station program and lead to the abrogation of several international commitments in space science and applications. (NASA Release 88-60).

May 12: The Associated Press and United Press International news agencies reported that U.S. surveillance satellites had detected the explosion of a Soviet solid propellant plant 500 miles southwest of Moscow in the Ukraine. The Soviet Union confirmed that an explosion had occurred—killing three people and seriously injuring five others—but denied that rocket propellants had been involved in the blast. U.S. intelligence officials claimed that the explosion destroyed several buildings at the plant and put a halt to all solid propellant manufacturing activities at the facility, which was the only source of solid propellant for the SS24 intercontinental ballistic missile (ICBM). Two weeks prior, a powerful series of explosions destroyed a solid propellant plant in Nevada, killing two plant employees and injuring hundreds of others. Pentagon officials estimated it would take at least six months for the Soviet facility to resume production of ICBM propellants. (*AP*, May 18/88; *UPI*, May 18/88; *W Post*, May 18/88; *WSJ*, May 18/88; *USA Today*, May 18/88; *W Times*, May 18/88)

May 13: NASA's Lewis Research Center and the NASA/Industry Advanced Turboprop Team were awarded the prestigious Collier Trophy at the annual Robert J. Collier Trophy Dinner hosted by the National Aviation Club. Lewis and the NASA/Industry Advanced Turboprop Team were honored for developing technology and testing advanced turboprop propulsion systems that offered dramatic reductions in fuel usage and operating costs for subsonic transport aircraft. (NASA Release 88-59)

May 14: An untended Soviet spacecraft carrying fuel, food, equipment, and mail, docked with the Mir space station. Two Soviet cosmonauts, on a long duration mission, tended Mir. The cosmonauts, Vladimir Titov and Musa Manarov, arrived at the space station in December and were expected to stay in space for about a year, breaking the 325–day endurance record set the previous year by Yuri Romanenko. (*FBIS-Sov-88-108*, Jun 6/88; *SSR 1988 038A;* *AP*, May 14/88)

May 17: A European Space Agency (ESA) Ariane 2 launch vehicle was successfully launched at 7:58 p.m. EDT from the Kourou Launch Center in French Guiana. The 162-foot, three-stage rocket successfully carried into orbit the 2,619-pound Intelsat 5 commercial communications satellite which would eventually propel itself into a 22,300-mile geosynchronous orbit. This was the 22nd launch of the Ariane series and the second Ariane 2 launch in 1988. A total of eight Ariane 2 launches were scheduled for 1988. (*SSR 1988 040A; AP*, May 18/88; *UPI*, May 18/88)

May 19: Science magazine published findings from a recent Earth-based radar mapping of the planet Venus, conducted by Jet Propulsion Laboratory (JPL) scientists. Newly processed radar pictures of the surface of Venus showed what may be geologically recent volcanic activity and impact cratering, including possible volcanic mountains and broad lava flows and fields like those which make up the Hawaiian Islands. The JPL radar images, with a resolution of less than a mile, were the latest in a series of Earth-based radar scans of Venus conducted at the Goldstone, California, complex of NASA's Deep Space Network. The Magellan spacecraft was expected to make more detailed radar observations of Venus in 1990. (NASA Release 88-65; *C Trib*, May 21/88)

May 21: In an interview with the *Washington Post,* Soviet leader Mikhail Gorbachev announced he would ask President Ronald Reagan to approve a joint Soviet-U.S. untended flight to Mars during the upcoming summit meeting in Moscow. A Soviet Foreign Ministry spokesman said Gorbachev's proposal for a Mars mission was recommended by Roald Z. Sagdeev, a Soviet space expert, as a symbol of superpower cooperation and a fruitful scientific and technological undertaking. "This is a field for cooperation that would be worthy of the Soviet and American people," Gorbachev said. "And I will make that proposal to President Reagan." (*W Post,* May 22/88; *UPI,* May 22/88; *LA Times,* May 22/88; *B Sun,* May 23; *C Trib,* May 23/88)

May 23: NASA and the Department of Commerce awarded the first U.S. Government contract for commercial launch transportation services to a private firm. The two agencies contracted General Dynamics Space Systems Division, San Diego, California, to provide expendable launch vehicle transportation services for the Department's National Oceanic and Atmospheric Administration (NOAA) Geostationary Operational Environmental Satellites (GOES). This contract marked the first time a contractor assumed systems performance responsibility for overall program and subcontractor management, vehicle design, production, testing and vehicle-to-payload integration, mission integration, launch services, system effectiveness, overall launch vehicle performance, and mission success.

In a related matter, NASA's Kennedy Space Center recently entered into an agreement with General Dynamics that allowed the company to use NASA Launch Complex 36 and associated facilities for commercial launch operations of the Atlas/Centaur rocket. (NASA Release 88-66; *UPI,* May 23/88; *AP,* May 24/88; *B Sun,* May 24/88; *NY Times,* May 25/88; *C Trib,* May 26/88)

• Negotiators from NASA and the government of Japan reached agreement in substance on the text of a memorandum of understanding (MOU) for cooperation in the detailed design, development, operation, and utilization of the permanently crew-tended civil Space Station. The Japanese Science and Technology Agency would serve as NASA's counterpart in implementing this MOU.

Under the terms of the new MOU, Japan would provide the Japanese Experiment Module (JEM)—a permanently attached, pressurized laboratory module, which would include an exposed facility and an experiment logistics module. The pressurized portion of the JEM would provide a shirt-sleeve environment for the Space Station crew to perform research activities. The JEM's exposed facility would be used for scientific observations, Earth observation, communications, advanced technology development and other activities requiring direct exposure to space. The experiment logistics module, which would provide transportation and storage of logistics items, would be transported to the Station by the Space Shuttle. (NASA Release 88-70)

June

June 6: Morton Thiokol announced that it would not bid on a $1.5 billion contract to build the proposed Space Shuttle Advanced Solid Rocket Motor (ASRM). Instead, Thiokol officials stated they would concentrate on building the current generation of redesigned solid rocket motors, currently undergoing final testing prior to the resumption of Space Shuttle flights in September. (*UPI,* Jun 6/88; *AP,* Jun 6/88; *NY Times,* Jun 7/88; *WSJ,* Jun 7/88; *LA Times,* Jun 7/88)

June 7: A Soviet Proton launch vehicle, carrying a Soyuz TM-5 spacecraft with a three-man Soviet-Bulgarian crew, took off from the Baikonur launch facility. The Soyuz TM-5 crew was to join a two-man crew aboard the Mir space station. The launch, which occurred at 10:04 a.m. EDT, was the first Soviet crew-tended launch in 1988. The Mir visit was to be a 10-day mission in which the visiting astronauts would perform over 40 experiments aboard the station. The two astronauts aboard Mir were to remain after the departure of their colleagues and continue a year-long endurance mission that was expected to break the record established by Yuri Romanenko in 1987. (FBIS-Sov-88-110, Jun 8/88; S*SR 1988 048B*; *NY Times,* Jun 8/88; *B Sun,* Jun 8/88)

June 8: Negotiations were completed among the United States, Canada, Europe, and Japan on the framework for international cooperation in the Space Station program.

Under the agreements, the United States would provide the overall Space Station framework: operating subsystems, including life support and 75 kilowatts of power; laboratory and habitat modules; and a free-flying platform that would be placed in polar orbit for Earth observation. Canada would provide a Mobile Servicing System, which would be used in conjunction with the assembly, maintenance, and servicing of Space Station elements. Japan would provide the Japanese Experiment Module, a permanently attached pressurized laboratory module, which would include an exposed facility and an experiment logistics module. The European Space Agency would provide a pressurized laboratory module, which would be permanently attached to the crew-tended base; a free-flying polar platform to work together with the U.S. polar platform; and an astronaut-tended free flyer to be serviced at the crew-tended base.

The United States anticipated spending approximately $16 billion (fiscal year 1989 dollars) to develop Space Station hardware. The total foreign commitment to the Space Station was in excess of $7 billion. Furthermore, the partners would cover more than 25 percent of the Space Station's expected annual operating costs throughout the 20–30 year life of the program. Signature of the agreements was expected later in the summer. (NASA Release 88-74)

June 9: The first direct observation of an atmosphere on Pluto was made by Massachusetts Institute of Technology (MIT) astronomers flying aboard NASA ARC's Kuiper Airborne Observatory (KAO). A team of astronomers from MIT, including Edward Dunham, James Elliot, Amanda Bosh, Steve Slivan, and Leslie Young, made the observation during a temporary occultation of a star behind Pluto. The observations were made at 41,000 feet (12,300 meters) altitude, approximately 500 miles south of Pago Pago, American Samoa, over the Southern Pacific Ocean. Information about the temperature, pressure, and extent of the atmosphere would be derived from the occultation data, obtained using a solid state video camera attached to KAO's 36 inch (92 cm) telescope. The airborne observations lasted about 1.5 minutes, occurring shortly after midnight. (ARC Release 88-40)

June 13: This date marked five years since the Pioneer 10 spacecraft had left the solar system on its trajectory toward interstellar space. Pioneer 10—the most distant human-made object in existence—was now 4,175,500,000 miles from the Sun, almost 45 times the distance from the Earth to the Sun. Radio signals, moving at the speed of light, took 12 hours and 26 minutes to travel from Earth to spacecraft and back, the longest time of any radio communication in history.

The spacecraft, launched in 1972, continued to operate extremely well as it collected and transmitted data back to Earth. Its primary mission, originally scheduled for 21 months, was to assess the feasibility of passage through the Asteroid Belt and provide the first close-up examination of *Jupi*ter and its moons. Pioneer 10 accomplished all of its original goals by December 1973. At that point, the mission was indefinitely extended. Scientists reprogrammed the probe to explore the Sun's atmosphere and to look for a tenth planet and gravity waves in the far outer solar system and beyond.

Recent improvements in the NASA ground stations were expected to allow communications with Pioneer 10 to continue until the range approached six billion miles, more than twice the pre-launch estimates. Project Manager Richard O. Fimmel expected that NASA would be able to track Pioneer 10 until the power source limited communications, which was expected to occur toward the end of the 1990s.

Both Pioneer 10 and its sister spacecraft, Pioneer 11, carried an easily interpreted graphic message in the event that they encountered any intelligent life forms on their journey. Scientists believed Pioneer 10 and 11 would travel among the stars virtually forever because the vacuum of interstellar space is so empty that the risk of any type of collision would be negligible. (ARC Release 88-36)

June 14: The fourth full-duration test firing of NASA's redesigned Space Shuttle solid rocket motor was conducted at Morton Thiokol's Space Operations facility near Brigham City, Utah. The test was part of the Shuttle motor redesign program. The 126-foot-long, 1.2-million-pound motor, designated Qualification Motor-7 (QM-7), underwent a full-duration, horizontal test firing of two minutes. After a preliminary examination of the fired motor, NASA and Morton Thiokol officials declared the test an apparent success. The exercise involved the first use of a new test stand at Morton Thiokol. One remaining full-duration test, Production Verification Motor-1 (PVM-1), would be conducted in July, prior to the next Space Shuttle flight. (NASA Release 88-7; *NY Times,* Jun 15/88)

June 16: An advanced Navy navigation satellite, NOVA-II, was launched into polar orbit on a NASA Scout launch vehicle from Western Space and Missile Center (WSMC), California. This was the 111th flight of the 73-foot long, four-stage, solid propellant Scout launch vehicles. (SSR 1988 052C; NASA Release 88-75)

June 26: This date marked the tenth anniversary of the launch of NASA's Seasat Satellite, which ushered in a new era in space research focusing on unsolved questions of the world's oceans and weather. Launched on June 26, 1978, on an Atlas-Agena rocket from Vandenberg Air Force Base, California, Seasat carried a payload of five scientific instruments unlike any package carried on previous remote-sensing satellites.

Among the experimental instruments Seasat pioneered were a synthetic aperture radar, which provided highly detailed images of ocean and land surfaces; a radar scatterometer to measure near-surface wind speed and direction; a radar altimeter to measure the height of the ocean surface and waves; and a scanning multichannel microwave radiometer to measure surface temperature, wind speeds, and sea ice cover. The satellite also carried a passive visual and infrared radiometer to provide supporting data for the other four experiments.

Seasat demonstrated how space sensors could be used in oceanography, serving as a protoype for a generation of international missions that could provide answers to some of the world's most baffling and threatening weather phenomena. (NASA Release 88-84)

June 28: Rear Admiral Richard H. Truly, NASA Associate Administrator for Space Flight, updated the target for the upcoming Space Shuttle launch to early September and indicated that he was very pleased with the progress of preparations for STS-26. The new launch target was selected following a major review covering all aspects of the Shuttle program and Truly's subsequent report to NASA Administrator Dr. James C. Fletcher. Truly said, "Hard work by a lot of people is paying off and the Shuttle program is coming along nicely."

Discovery was scheduled to be rolled out to the Kennedy Space Center Launch Pad 39B the first week of July. A flight readiness firing (FRF) of Discovery's main engines and liquid propulsion system was targeted for late July. The final full-scale, full-duration test firing of the redesigned Shuttle solid rocket motor (SRM) prior to STS-26 was also planned for late July. (NASA Release 88-85)

June 30: The NASA Administrator appointed Ivan Bekey to be Special Assistant for System Engineering and Planning Integration, Office of Exploration. Bekey would be responsible for overall systems engineering for Office of Exploration studies and formulation of top level architecture. In the previous year, he served as Director of Program Planning for the Office of Policy and Planning. For the preceding six years, he was the Director of Advanced Programs, Office of Space Flight. He was also NASA coordinator for the National Commission on Space. Bekey joined NASA in 1978. (NASA Special Announcement on 06-60-88)

July

July 1: NASA officially transferred custody of Launch Complex 17 and East Coast Delta launch operations to the U.S. Air Force. Since 1960, NASA had conducted 143 Delta launches from the two-pad Complex, located at Cape Canaveral Air Force Station, Florida. Under Air Force stewardship, Complex 17 would continue to be used to launch Delta medium class vehicles. The Air Force had procured 20 new Delta IIs for Department of Defense payloads. The first launch was scheduled for later in 1988. In addition, at least eight commercial Delta IIs would be launched by McDonnell Douglas Astronautics Company, Huntington, Beach, California, from Complex 17 between 1989 and 1992. (NASA Release 88-99)

July 1: NASA's Jet Propulsion Laboratory (JPL), Pasadena, California, inaugurated its Mark III Hypercube parallel supercomputer, the result of a five-year research and development effort at the JPL Center for Space Microelectronics Technology and the California Institute of Technology (Caltech). The Mark III spearheaded the arrival of massively parallel super-computing. Its first module, placed on-line at the Caltech computer network, contained 32 nodes, or processing units, which together had a peak speed of about 512 million floating point operations per second. Three more 32-node modules would be added during the next nine months to form a 128-node hypercube with a peak speed of 2 billion floating point operations per second.

The Mark II Hypercube was being used for scientific, engineering, and defense research applications, including analysis of NASA multispectral space imaging data; analysis of NASA synthetic aperture radar images taken from the Space Shuttle; and simulation of the Strategic Defense System. (NASA Release 88-88)

July 4: The fully assembled Space Shuttle Discovery vehicle was rolled out of the Kennedy Space Center Vehicle Assembly Building (VAB) to Launch Pad 39-B in preparation for mission STS-26, scheduled for late September. (*UPI*, Jul 4/88; NY Times, Jul 5/88; W Post, Jul 5/88; WSJ, Jul 5/88; W Times, Jul 5/88; P Inq, Jul 5/88)

July 5: An *ad hoc* committee, assembled to review safety risk management in the National Space Transportation System program, completed its report, finding that there had been "a positive change in attitudes" by NASA and NASA contractors towards safety. The six-member committee, composed of both Government and independent safety experts, found no significant system safety issues that would adversely impact the STS-26 launch. (NASA Release: 88-89)

July 7: The Soviet Union successfully launched the Phobos 1 spacecraft to Mars aboard a 198-foot Proton launch vehicle from the Baikonur launch facility. Phobos 1 and Phobos 2, scheduled for launch the following week, would carry 100-pound landers designed to analyze Phobos, the larger of the two moons of Mars. The 17-mile-long, potato-shaped Phobos and its smaller moon Deimos were thought to be asteroids captured by the gravitational field of Mars. Tracking support for the Soviet missions would be provided by NASA's Deep Space Network (DSN), whose primary mission would be to provide essential tracking data to permit landings on Phobos. The DSN would then shift to enabling a key scientific goal of the mission, to track Phobos very precisely. The 13,000-pound Phobos spacecraft would carry more than two dozen robotic instruments on their 200-day journey to Mars. (FBIS-Sov-88-131, Jul 8/88; SSR 1988 058A; NASA Release 88-87; *UPI*, Jul 8/88; *NY Times*, Jul 8/88; *W Post*, Jul 8/88;)

July 12: The Soviet Union successfully launched the Phobos 2 spacecraft to Mars aboard a 198-foot Proton launch vehicle from the Baikonur launch facility. The Phobos 2 launch occurred less than a week after the launch of its companion spacecraft, Phobos 1. (FBIS-Sov-88-134, Jul 13/88; SSR 1988 059A; *W Post*, Jul 13/88; *W Times*, Jul 13/88; *LA Times*, Jul 13/88; *P Inq*, Jul 13/88; *C Trib*, Jul 13/88)

• NASA announced that a scientist at the Ames Research Center, Mountain View, California, had developed a new, ground breaking computational model that could accurately simulate the complex, fluctuating airflow within aircraft engine turbines and compressors. The new computational model, developed by NASA scientist, Man Rohan Rai, was expected to generate significant savings for the aircraft engine industry once it became ready for commercial applications. The model performed one of the most complex computer simulations ever undertaken and provided the most accurate calculation to date of air flow within turbines. Engine concepts and designs had been tested primarily by experimental methods, most often by building prototype engines and subjecting them to a battery of tests. Advanced computer techniques were far less expensive and much faster. Current computational work was performed on a Cray-2 supercomputer at the new Ames Numerical Aerodynamics Simulation Facility. (NASA Release 88-94; ARC Release 88-50; *SF Chron*, Jul 13/88; *Bus Week*, Jul 25/88)

July 13: In a setback for the Space Station program, the Senate approved a $10.1 billion appropriation for fiscal year 1989—$1.4 billion less than was requested by NASA and the Reagan administration. The legislation provided only $200 million of the estimated $767.4 million NASA said it needed to continue basic work on the Space Station program. An earlier House appropriations bill approved $902 million of the $967 million requested for the Station. Senior Reagan administration officials warned that the final version of the bill would provoke a Presidential veto if it lacked sufficient funding to continue the Space Station program. (*UPI*, Jul 13/88; *WSJ*, Jul 14/88; *W Times*, Jul 14/88; *USA Today*, Jul 14/88; *LA Times*, Jul 14/88)

• An Indian-built rocket carrying a weather satellite malfunctioned and plunged into the Indian Ocean shortly after it was launched from India's Sriharikota Island launch facility. The five-stage, 80-foot rocket Augmented Satellite Launch Vehicle (ASLV) apparently suffered a malfunction in the solid fueled first stage boosters. This was the second failure of the new rocket, which suffered a catastrophic malfunction during its first launch attempt in March 1987. The launch failure was interpreted by space experts as a major setback for India's civilian space and ballistic missile development programs. (*NY Times*, Jul 14/88; *C Trib*, Jul 14/88; *P Inq*, Jul 14/88; *LA Times*, Jul 14/88)

July 13: The Administrator appointed Edward A. Frankle to be NASA general counsel. Frankle had been Deputy General Counsel since October 27, 1985. He began his career at NASA in September 1982, as Chief Counsel of the Goddard Space Flight Center, Greenbelt, Maryland. (NASA Release 88-98)

July 15: The launch of the Space Shuttle Discovery faced further delays as a result of the discovery of a fuel leak in a fuel line inside one of the Shuttle's orbital maneuvering system (OMS) pods. NASA engineers determined that the small leak, which was detected during a test of the OMS fuel tanks, would have to be fixed prior to the launch of Discovery. The repair might delay the Shuttle launch by up to two months, depending on whether the spacecraft could be serviced at the launch pad or would have to be rolled back to the vehicle assembly building at Kennedy Space Center. NASA officials said the leak would not affect the scheduled test firing of the Shuttle main engines on July 28. (*NY Times*, Jul 18/88; *W Post*, Jul 18/88; *W Times*, Jul 18/88; *USA Today*, Jul 18/88; *P Inq*, Jul 18/88)

July 18: President Ronald Reagan christened the planned international Space Station "Freedom," to symbolize the values shared by the U.S. and its allies. White House spokesman Marlin Fitzwater said the name "Freedom" was recommended by NASA officials and representatives of Canada, Japan, and the European Space Agency, project participants. (WH Release, Jul 18/88; *UPI*, Jul 18/88; *W Post*, Jul 18/88; *W Times*, Jul 18/88)

July 20: NASA announced it would test—throughout the month—a new deep-space communications system with its Voyager 2 spacecraft, in preparation for the Voyager 2 fly-by of Neptune in August 1989. As part of the new system, NASA added the 27 telescopes from the National Radio Astronomy Observatory's (NRAO) Very Large Array (VLA) facility to the Jet Propulsion Laboratory's deep space network (DSN). This action more than doubled DSN's ability to capture the Voyager signal, which became extremely faint as it approached the vicinity of Neptune. Under an agreement between NASA and the National Science Foundation, NRAO sponsors, engineers installed new receivers and microwave horns, tuned to the Voyager X-Band radio frequency, on all the 82-foot dish antennas at the VLA. Special signal-processing and communication equipment was added so that the VLA would be linked by satellite to the DSN's Deep Space Communication's Complex at Goldstone, California. (NASA Release 88-102)

July 21: An ESA Ariane 3 rocket carrying two communications satellites was successfully launched from the Korou Launch Center in French Guiana. The Ariane 3, launched at 7:12 p.m. EDT, carried two satellites belonging to India, and the European Telecommunications Satellite (Eutelsat), into polar orbit. (*UPI,* Jul 22/88; *W Times,* Jul 22/88)

July 26: NASA announced selection of the Government sites to be used as production and testing facilities for the Agency's planned Space Shuttle Advanced Solid Rocket Motor (ASRM). The Tennessee Valley Authority property known as Yellow Creek, in northeastern Mississippi, was selected as the location of the ASRM production facility, while NASA's Stennis Space Center near Bay St. Louis, Mississippi, was selected for ASRM testing. Maximum utilization also would be made of available manufacturing and computer capability at the Michoud Assembly Facility and the Slidell Computer Complex, both located in southeast Louisiana, to minimize total program costs.   The planned ASRM, which would replace the current Redesigned Solid Rocket Motors in the mid-1990s, would incorporate substantive design changes to improve reliability and safety margins, as well as provide a significant added performance capability to the Space Shuttle. (NASA Release 88-104; MSFC Release 88-102; *UPI,* Jul 27/88; *W Post,* Jul 27/88; *USA Today,* Jul 27/88)

During July: A team of NASA scientists from the Marshall Space Flight Center in Huntsville, Alabama, the Lockheed Company, and the University of Alabama in Huntsville, announced the discovery of a new high-temperature superconductor. According to a paper published in *Applied Physics Letters,* the scientists demonstrated that "samples of yttrium-barium-copper oxide, when mixed with silver oxide, heat treated, and exposed to -320 degrees Fahrenheit, can be suspended below a rare Earth magnet by the magnetic field trapped in the sample." Because superconducting material excludes a magnetic field, by floating a superconductor beneath a magnet, the two materials are firmly attracted to each other, but never touch. According to Palmer N. Peters, of Marshall's Cryogenics Physics Office and a member of the scientific team, the implications of high temperature superconductors for space technology are numerous. "In space, with the elimination of the weight associated with gravity, it should be possible to develop low-vibration, low-friction couplings and bearings," Peters said. "The importance of the discovery is that this new material not only has stronger suspension forces but exhibits other unusual magnetic properties: it has demonstrated a lower electrical resistance at normal temperatures, is easier to solder, and is less brittle than other high temperature superconducting materials." (MSFC Release 88-104)

August

August 1: A Wet Countdown Demonstration Test (WCDT) of the Space Shuttle Discovery's External Tank (ET) was conducted following several unplanned holds as a result of problems with ground support equipment. The WCDDT, carried out with Discovery deployed for launch at the Kennedy Space Center's Launch Complex 39-B, consisted of loading the ET with liquid Oxygen and Liquid Hydrogen in mock preparation for a Shuttle launch. (*NASA STS-26 Press Kit*)

August 2: NASA launched the first of three untended balloons that would search for cosmic rays, including those that could provide evidence of galaxies made of antimatter. The balloon flights were to be conducted throughout August from a launching site at Prince Albert Saskatchewan Airport, Canada, as part of the NASA Balloon Program. Each of the balloons would carry one of three cosmic ray experiments: an extragalactic matter experiment that would search for heavy antinuclei; a scintillating optic fiber experiment; and a high energy spectrometer telescope. (NASA Release 88-106; GSFC-WFF Release 88-25)

August 3: NASA held dedication ceremonies for the newly renamed John C. Stennis Space Center—formerly the National Space Technology Laboratories—in Hancock County, Mississippi. The Center was named for Mississippi's retiring senator by an Executive order signed May 20, 1988, by President Ronald Reagan. (NASA Release 88-93)

August 4: NASA appointed Gary L. Tesch to be NASA Deputy General Counsel. Tesch had been serving as Associate General Counsel (contracts) prior to this appointment. (NASA Release 88-109)

August 8: Astronomers at NASA's Space Telescope Science Institute, Baltimore, Maryland, and the University of California at Berkeley uncovered the most distant galaxy yet seen. Called 4C41.17, the newly discovered galaxy is located at an estimated distance of 15 billion light years—more than 90 percent of the distance to the visible limits of the universe. The discovery was made by a team of American and European astronomers, using the Very Large Array (VLA) Radio Telescope Facility near Socorro, New Mexico, and was confirmed a galaxy using the 2.1 meter optical telescope at Kitt Peak National Observatory. Studying remote galaxies like 4C41.17 was expected to help forge a better understanding of how galaxies have evolved since the time of the Big Bang. This research was supported by NASA, the European Space

Agency, the National Science Foundation, the Hopkins Ultraviolet Telescope project, and the Space Telescope Science Institute. (NASA Release 88-111; *P Inq*, Aug 9/88; *B Sun*, Aug 9/88)

August 10: After an aborted first attempt on August 4, a 22-second flight readiness firing (FRF) of the Space Shuttle Discovery's main engines was conducted on the launch pad at the Kennedy Space Center's Launch Complex 39-B. The first FRF attempt had been halted inside the T-10 second mark because of a sluggish fuel bleed valve on the number 2 main engine. The faulty valve was replaced prior to the FRF. (*AP*, Aug 10/88; *UPI*, Aug 10/88; *NY Times*, Aug 11/88; *W Post*, Aug 11/88; *WSJ*, Aug 11/88; *CSM*, Aug 11/88; *W Times*, Aug 11/88; *B Sun*, Aug 11/88; *P Inq*, Aug 11/88)

August 11: A Space Systems Development Agreement was signed by NASA and Spacehab, Inc., a private firm that was developing a Shuttle-based pressurized module for commercial applications. According to the agreement, NASA agreed to provide Shuttle payload bay space for Spacehab's middeck augmentation module on six Shuttle flights beginning in 1991. Spacehab would pay $28.2 million (adjusted for inflation) for standard services for each flight. The Spacehab module, to be constructed by McDonnell Douglas Astronautics Company, Huntsville, Alabama, would be a metal truncated cylinder measuring 10 feet in length by 13 feet in diameter, designed to fit in the Shuttle's cargo bay. Shuttle crew members would access the module through a tunnel from the Shuttle's middeck. The company would make available to customers a variety of locker and rack accommodations, with associated support and integration services. Spacehab was specifically identified in the President's Commercial Space Initiative, announced in February, in which the Reagan administration committed to make the best efforts to launch the commercial module in the early 1990s.(NASA Release 88-114)

August 18: The fifth and final test of NASA's redesigned Space Shuttle solid rocket motor, prior to resumption of Shuttle flights, was conducted at Morton Thiokol's Space Operations facility near Brigham City, Utah. The 126-foot-long, 1.2 million-pound motor, designated Production Verification Motor-1 (PVM-1), underwent a full duration, horizontal test firing of two minutes. The motor was extensively flawed to demonstrate the fail-safe characteristics of the redesign. (NASA Release 88-113; *AP*, Aug 18/88; *UPI*, Aug 18/88; *NY Times*, Aug 19/88; *W Post*, Aug 19/88; *WSJ*, Aug 19/88; *USA Today*, Aug 19/88; *W Times*, Aug 19/88; *B Sun*, Aug 19/88)

• NASA announced that TRW, Inc. had been selected for final negotiations leading to the award of contracts for extended definition and development of the space-based Advanced X-ray Astrophysics Facility (AXAF). The facility would be the third in NASA's series of space-based great observatories, following the Hubble Space Telescope and the Gamma Ray Observatory into orbit in the mid-1990s. These observatories, as well as the Space Infrared Telescope Facility, would permit simultaneous, complementary observations of astrophysical phenomena over different wavelengths of the spectrum. The objective of this project would be to develop a high-quality, x-ray telescope to be used by the international scientific community, in conjunction with NASA, for an operational period of 15 years. The proposed cost of the contract was approximately $508 million. (NASA Release 88-118; MSFC Release 88-131)

August 25: NASA launched two U.S. Navy navigational satellites aboard a Scout rocket from Vandenberg Air Force Base, California. The two Oscar satellites, each weighing 141 pounds, were placed into a 600-nautical-mile circular polar orbit. The Oscars were part of the Navy's all-weather, global navigation system. The Stacked Oscars on Scout-4 (SOOS-4) mission marked the fourth and final Scout launch for 1988. (SSR 1988 074A-G; NASA Release 88-116)

August 29: A Soviet Proton launch vehicle carrying a Soyuz TM-6 spacecraft, with three cosmonauts on board, was successfully launched from the Baikonur Cosmodrome. The three-man Soyuz crew, which included an Afghan and a Soviet physician, was scheduled to join a two-man crew aboard the Mir space station on the week-long mission. The physician, Valery Polyakov, was to remain aboard Mir after the departure of his Soyuz crewmates and continue monitoring the health of the Mir cosmonauts, who were in the eighth month of a record-breaking year-long endurance mission. (FBIS-Sov-88-168, Aug 30/88; SSR 1988 075A; *UPI,* Aug 30/88; *W Post,* Aug 30/88; *CSM,* Aug 30/88; *P Inq,* Aug 30/88)

August 30: NASA issued the newest update of its mixed-fleet manifest, reflecting current planning for primary payloads for Space Shuttle missions and expendable launch vehicles (launch vehicles) through fiscal year 1993. The manifest reflected changes made to accommodate a slower than expected resumption of Shuttle flights and the necessity of rescheduling to conserve the launch windows for three major planetary missions. Among the major changes made was a rescheduling of the Hubble Space Telescope launch from June 1989 to February 1990 and the transferral of a Department of Defense Shuttle mission to an expendable launch vehicle. (Payload Flight Assignments-NASA Mixed Fleet, Aug 1988; NASA Release 88-120)

During August: A Cray Y-MP, the fastest supercomputer in the world, arrived at NASA's Ames Research Center, Mountain View, California, and underwent acceptance tests in the Numerical Aerodynamic Simulation (NAS) Facility. The NAS was established as a national facility available to leading U.S. research institutions through a national network. It had long-term support from Congress to use the fastest available computers to achieve rapid progress in U.S. aerospace work and in other fields such as weather, astrophysics, and chemistry.

Supercomputer systems were complementing and extending aerospace research being done in wind tunnels and high-velocity (4,000 to 15,000 mph) shock tunnels. The new Y-MP central processor would not only solve old problems four times as fast as existing systems, but also would be used to investigate flow phenomena that could not be investigated currently by wind tunnels or computers. Another important application of the NAS supercomputer system would be the modeling of airflows for hypersonic vehicles such as the National Aero-Space Plane and the Space Shuttle.

Ames was the first customer to take delivery of this new supercomputer, expected to be fully operational by January 1989. (ARC Release 88-80)

September

September 1: The fifth firing in a series of Transient Pressure Test Articles (TPTA) was conducted in the East Test Area of NASA's Marshall Space Flight Center in Huntsville, Alabama. The test, designated TPTA 1.3, was part of the Space Shuttle solid rocket motor redesign program. For this test, flaws were intentionally cut into both the test article's field joints, the case-to-nozzle joint, and other components of the redesigned motor. Instrumentation on the motor recovered data to verify the structural performance, thermal response, and sealing capability of the redesigned field and case-to-nozzle joints. (MSFC Release on 08/30/88)

September 2: A classified Department of Defense payload was launched aboard a Titan 34D launch vehicle from launch complex 40 at Cape Canaveral Air Force Station. Sources quoted by the Associated Press said the rocket failed to carry its payload into the correct orbit. The payload was widely believed to be a Vortex reconnaissance satellite. (SSR 1988 077A-C; *UPI*, Sep 2/88; *Reuter*, Sep 2/88)

September 5: A classified Department of Defense payload was successfully launched aboard a Titan 2 launch vehicle from Vandenberg Air Force Base, California. Civilian experts said they believed the rocket was carrying a cluster of Navy recon- naissance satellites known as "White Cloud." The Titan 2 launch followed an apparent failure of a Titan 34D rocket to launch a military reconnaissance satellite into proper orbit. (*SSR 1988 078A-B; NY Times,* Sep 6/88; *W Times,* Sep 6/88; *CSM,* Sep 6/88)

September 6: China successfully launched its first experimental weather satellite, the Wind and Cloud No. 1. The satellite was launched by a Long March 4 launch vehicle from the Taiyuan launch facility in north-central China. (FBIS-CHI-88-90, Sep 16/88; SSR 1988 080A-B; *AP,* Sep 7/88; *W Post,* Sep 8/88)

September 7: After a tense 24-hour delay of deorbit burn, a Soviet Soyuz TM-5 capsule carrying two cosmonauts landed safely in Soviet Central Asia. The Soyuz TM-5, returning to Earth from a week-long visit to the Mir space station, experienced premature shutdowns of its reentry motor during the first two deorbit burn attempts. The first shutdown had been caused by a malfunction in the automatic guidance system, whereas the second failure resulted from improper resetting of the flight computer. The problem was considered by some U.S. space experts as potentially life threatening because the Soyuz was estimated to have only 24 to 48 hours of oxygen left and a limited amount of fuel with which to attempt a third deorbit burn. The mission had been noted for the participation of an Afghani cosmonaut and the transfer to the Mir space station of a physician crew member. (FBIS-SOV-88-175, Sep 9/88; *AP,* Sep 7/88; *NY Times,* Sep 7/88; *W Post,* Sep 7/88; *WSJ,* Sep 7/88; *W Times,* Sep 7/88; *B Sun,* Sep 7/88)

• NASA and McDonnell Douglas Astronautics Company, St. Louis, Missouri, announced the signing of an agreement providing for use of facilities at the Kennedy Space Center, Florida, and technical support from the Goddard Space Flight Center, Greenbelt, Maryland, in support of commercial launches. The umbrella agreement would enable McDonnell Douglas to gain access to NASA-managed launch support facilities when conducting commercial launches of the Delta rocket, expected to begin in 1989. (NASA Release 88-124)

September 8: An ESA Ariane 3 launch vehicle carrying two U.S. communications satellites was launched from the Korou launch facility in French Guiana. The Ariane 3 was carrying the 1.3-ton G-STAR 111 and 1.2-ton SBS 5 telecommunications satellites, both designed for television, telephone, and digital data transmission. Upon reaching low-Earth orbit, the G-STAR satellite failed to boost itself into its planned 23,000-mile geosynchronous orbit. (FBIS-Lat-8-175, Sep 9/88; SSR 1998 081C-D; *AP,* Sep 7/88; *Reuter,* Sep 7/88; *W Post,* Sep 9/88; *NY Times,* Sep 15/88; *WSJ,* Sep 15/88; *W Post,* Sep 15/88)

• A guidance system malfunction caused the Soviet Mars observer spacecraft Phobos 1 to wobble out of control on its way toward Mars. The spacecraft stopped responding to commands from Soviet ground controllers and was rapidly losing power as it was unable to properly orient its solar panels toward the Sun. Phobos 2 continued to function normally. (*UPI,* Sep 9/88; *NY Times,* Sep 9/88; *W Times,* Sep 9/88)

September 10: An untended Soviet Progress-38 supply spacecraft, bound for the Mir space station, was launched from the Baikonur launch facility. The Tass News Agency reported the Progress-38 ferried "expendable materials and various loads" for use by the Mir three-man crew. (SSR 1988 083A; *C Trib,* Sep 11/88)

September 12: NASA and the Air Force, responsible for range safety, mutually determined that the number of persons permitted access to the Kennedy Space Center, Florida, for the launch of the Space Shuttle Discovery, STS-26, would be greatly reduced for safety reasons. Studies of the January 1986 Challenger accident, and the loss of an Air Force Titan 34D the following April, showed that the danger to persons on the ground, if an accident occurred, was much greater than previously thought. Under certain conditions, a solid rocket booster released from the vehicle as a result of an accident would follow an unknown ballistic trajectory, rather than tumble on course as previously thought. In addition, upon destruction the boosters fragmented into thousands of pieces and the explosive properties of unburned solid motor propellant were greater than originally determined in laboratory tests. Because of this new information, NASA and the Air Force determined that the prudent action would be to minimize the population at the close-in viewing areas. (NASA Release 88-125)

September 16: NASA set a target launch date of September 29 for STS-26, the next Space Shuttle flight. The establishment of the launch date followed an updated assessment of the projected impact of Hurricane Gilbert on mission control and training facilities at the Johnson Space Center (JSC), Houston, Texas. In connection with the announcement, Rear Admiral Richard H. Truly, NASA Associate Administrator for Space Flight, said, "NASA's decision to set this launch date is based on over two years of persistence and dedication by NASA and contractor personnel, culminating in the STS-26 flight readiness review held at the Kennedy Space Center on September 13 and 14. I'm delighted to have reached this point and my hat is off to all members of the Shuttle team whose tireless efforts have brought us here to the brink of America's return to manned spaceflight." (NASA Release 88-127)

September 19: Israel launched its first orbital satellite aboard an Israeli-built rocket. The Horizon I experimental satellite was launched from an Israeli launch facility in the Negev desert aboard what U.S. experts believed was either a Jericho II or a Comet launch vehicle. The Horizon I was expected to have an operational lifetime of one month in low-Earth orbit. (SSR 1988 087A-B; *NY Times,* Sep 20/88; *W Post,* Sep 20/88; *W Times,* Sep 20/88; *CSM,* Sep 20/88; *B Sun,* Sep 20/88; *LA Times,* Sep 20/88)

September 24: The National Oceanic and Atmospheric Administration's NOAA-11, meteorological weather satellite was launched aboard a U.S. Air Force Atlas-E launch vehicle from Launch Complex 3 at Vandenberg Air Force Base, California. NOAA-11, and its companion spacecraft, NOAA-9 and NOAA-10, assist in global and local weather forecasting. In addition, the NOAA satellites are used for hurricane tracking and warning; global sea ice monitoring; various atmospheric studies; and for agricultural, commercial fishing, forestry, maritime, and other industrial uses. NOAA-11 was placed in a 540-mile Sun-synchronous, near polar orbit, where it would circle the Earth approximately every 102 minutes, observe a different portion of the Earth's surface on each orbit, and view the Earth's entire surface and cloud cover once every 12 hours. (NASA Release 88-126; *Aeronautics and Space Report of the President: 1988;* SSR 1988 089A-C)

September 29–October 3: U.S. crew-assisted space flight resumed after a hiatus of over two years with the successful launch of the Space Shuttle

Discovery on mission STS-26 from Kennedy Space Center Launch Complex 39-B in Florida. The first Space Shuttle flight since the Challenger accident in January 1986, STS-26 followed a prolonged period of review and reassessment of the entire Shuttle program and redesign of the Shuttle's solid rocket motors (SRMs), which had experienced a catastrophic failure on STS-25. STS-26 had as its primary payload the Tracking and Data Relay Satellite-C (TDRS-C), which would significantly expand communications and data links between Earth and the orbiting Shuttles. In addition, Discovery carried the Orbiter Experiments Program Autonomous Supporting Instrumentation System (OASIS), which would record environmental data in the orbiter payload bay during STS flight phases, as well as 11 secondary payloads. The STS-26 mission lasted four days and concluded with a landing at Edwards Air Force Base, California, on October 3. The five-man crew for STS-26 included Navy Captain Frederick H. Hauck (Commander), Air Force Colonel Richard O. Covey (Pilot), and Mission Specialists John M. Lounge, Marine Lieutenant Colonel David C. Hilmers, and George D. Nelson. (*NASA STS-26 Press Kit*, SSR 1988 091A; *UPI*, Sep 29/88; *AP*, Sep 29/88; *Reuter*, Sep 29/88; *NY Times*, Sep 30/88; *W Post*, Sep 30/88; *W Times*, Sep 30/88; *USA Today*, Sep 30/88)

October

October 1: NASA marked its thirtieth anniversary with ceremonies at the Centers. (*Aeronautics and Space Report of the President: 1988*)

October 17: A localized electrical fire occurred onboard the Magellan spacecraft during a power systems check at the Kennedy Space Center's (KSC) Spacecraft Encapsulation Facility (SAEF-2). The fire, which occurred while a technician worked within the spacecraft, was quickly extinguished without injuries and with only minor damage to Magellan. The KSC fire department also responded but was not required to enter the SAEF-2 clean room environment. The incident was being investigated by NASA. (NASA Releases 88-141 and 88-153).

October 19: Dr. Noel W. Hinners was appointed Associate Deputy Administrator by NASA Administrator Dr. James C. Fletcher. Hinners had been serving as Associate Deputy Administrator. He joined NASA in 1972 as Deputy Director of Lunar Programs, Office of Space Science. From 1974 to 1979, he was NASA Associate Administrator for Space Science. He was Director of the National Air and Space Museum, Smithsonian Institution, prior to his 1982 appointment as Director of the Goddard Space Flight Center, Greenbelt, Maryland. (NASA Release 88-138)

• NASA Administrator Dr. James C. Fletcher received the International Technology Institute's (ITI) William F. Rockwell, Jr. Medal at the International Congress on Technology and Technology Exchange banquet, in Pittsburgh. The Rockwell Medal, awarded annually to a maximum of three individuals from different continents, recognizes contributions toward the generation, transfer, and application of technology for the betterment of mankind. (NASA Release 88-139)

October 20: Two of the three cosmonauts aboard the Mir space station took a four-hour spacewalk to replace a faulty component on an x-ray telescope attached to the station. Cosmonauts Vladimir Titov and Musa Manarov, on a year-long endurance mission aboard Mir, wore a new spacesuit design intended to increase flexibility. (*C Trib,* Oct 21/88)

October 25: NASA announced that it was rescheduling the launch of the Hubble Space Telescope from February 1990 to an earlier date of December 1989. The earlier date was made possible following reassessment of a variety of factors, including payload requirements and Space Shuttle orbiter assignments during the period. (NASA Release 88-143)

November

November 1: NASA's Goddard Space Flight Center, Greenbelt, Maryland, issued a request for a proposal that would lead to the award of a prime contract for the design, development, testing and fabrication of Space Station Freedom's flight telerobotic servicer (FTS). The FTS would be a space robot with automated features that would assist crews in the assembly, maintenance, and servicing of the Freedom station and visiting spacecraft. Mandated by Congress, the FTS program would consist of a developmental flight test on the Space Shuttle in 1991, followed by a demonstration test flight of the prototype robot on the Space Shuttle in 1993. The FTS was scheduled to be launched on the second Space Station Freedom assembly flight. (NASA Release 88-150)

• NASA's Lewis Research Center, Cleveland, Ohio, and the Department of Energy's Argon National Laboratory, Chicago, Illinois, signed an agreement to begin joint research in the development of high-temperature superconductivity (HTS) materials and technology. The objective of this major research effort would be to exploit recent rapid advances in HTS technology for significant space and aeronautical applications. Research and technology development would be undertaken in those areas where HTS could be an enabling technology or substantially improve existing systems.

Initially, research would concentrate on advanced studies and critical exploratory experiments to identify the most promising applications for further development. Among the first candidate applications would be superconducting magnetic energy storage, space electromagnetic propulsion, microwave power transmission, aeropropulsion, and electromagnetic launch systems. (NASA Release 88-149)

November 2: The Space Shuttle Atlantis was rolled out to Launch Pad 39B for a scheduled launch in late November of STS-27, a classified Department of Defense mission. (NASA Release 88-171)

November 6: A classified Department of Defense payload was successfully launched into orbit aboard a Titan 34D launch vehicle. (*Aeronautics and Space Report of the President: 1988*)

November 10: NASA announced that it had negotiated a contract with Rockwell International's Space Transportation Systems Division to consolidate

Space Shuttle orbiter logistics operations at the Kennedy Space Center (KSC), Florida launch site. Under the contract, Rockwell's orbiter logistics operations in Downey, California, would be transferred to existing Launch Support Operations activity at KSC. During this same period, Rockwell's component overhaul and repair activities would move from its network of original Shuttle equipment manufacturers to the Rockwell Service Center at Cape Canaveral, Florida. The move would complete a consolidation begun in 1985. (NASA Release 88-156)

November 11: Controllers at the NASA Jet Propulsion Laboratory, Pasadena, California, directed the Voyager 2 spacecraft to fire its hydrazine thrusters for 3 minutes and 29 seconds in order to steer the craft toward a fly-by of Neptune on August 28, 1989. The Voyager 2 trajectory would bring it to within 3,000 miles of Neptune's cloudtops. (*Voyager 2 Neptune Encounter Press Kit; LA Times,* Nov 14/88)

November 14: NASA Administrator Dr. James C. Fletcher appointed Dr. Franklin D. Martin Assistant Administrator for the Office of Exploration. Martin had been serving as the Deputy Associate Administrator for the Space Station office since September 1986. From 1983 through 1986, Martin was Director of Sciences and of Space and Earth Sciences at NASA's Goddard Space Flight Center, Greenbelt, Maryland. From 1974 through 1983, he served in a variety of positions at NASA Headquarters, including manager of the advanced programs in astrophysics, solar terrestrial, and lunar divisions. (NASA Release 88-158)

November 15: After several delays, the Soviet Union successfully completed an automated orbital flight test of the Space Shuttle Buran. The untended Soviet orbiter—which is almost identical in aerodynamic design and dimensions to the U.S. orbiters—was launched from the Baikonur cosmodrome in Central Asia. Buran was powered by a 198-foot Energia rocket with two strap-on boosters in a configuration similar to that of the U.S. Space Shuttle. Unlike the Shuttle, however, the Soviet orbiter does not carry its own main engines, but relies almost entirely on the Energia's engines for ascent to orbit. According to Soviet sources, the Soviet shuttle can be equipped with jet engines to assist landing—a capability the U.S. Shuttles lack. After completing two orbits, Buran landed successfully on a 2.7-mile runway at the Baikonur facility, ending the 3-1/2 hour mission. The Tass News Agency hailed the Buran mission as a "major success" and as a prelude to crew-tended Shuttle flights. In Washington, NASA Administrator Dr. James C. Fletcher congratulated the Soviet Union. (FBIS-SOV-88-222, Nov 17/88; SSR 1988 100A; *AP,* Nov 15/88; *UPI,* Nov 15/88; *NY Times,* Nov 16/88; *WSJ,* Nov 16/88; *W Post,* Nov 16/88; *LA Times,* Nov 15/88; *W Times,* Nov 16/88; *B Sun,* Nov 16/88; *P Inq,* Nov 16/88; *C Trib,* Nov 15/88)

• In a major setback for the U.S. radio astronomy program, the National Radio Astronomy Observatory's (NRAO) 300-foot radiotelescope in Greenbank, West Virginia, unexpectedly collapsed, completely destroying the metallic antenna dish. The instrument, built in 1962, was one of the most powerful radiotelescopes in the world and had been responsible for several important discoveries, e.g. pulsars. The cause of the collapse was not known, but was not believed to be related to wind or other atmospheric conditions. The NRAO Assistant Director George Seielsand noted that the telescope structure, originally designed for a dish diameter of 140-feet, had been expanded to a diameter of 300 feet without a corresponding increase in material strength. No decision was announced on whether funds would be sought for a replacement radiotelescope. (*NY Times,* Nov 17/88; *W Post,* Nov 17/88; *B Sun,* Nov 17/88)

November 21: NASA Administrator Dr. Fletcher appointed Kenneth S. Pedersen to be Associate Administrator for External Relations, effective immediately. Pedersen would be responsible for congressional, intergovernmental, international, and industry relations, educational activities and relations with educational institutions, and the NASA History Office. (NASA Release 88-160)

November 23: NASA announced the selection of General Electric, Astro-Space Division, Princeton, New Jersey, for negotiations leading to the award of a cost-plus-award-fee contract for design, fabrication, instrument integration, and launch operation support of the Global Geospace Science (GGS) Wind and Polar missions. The two GGS laboratories, with their complement of scientific instruments, were to examine the flow of energy from the Sun through the Earth's geospace environment. This project was to be part of the overall scientific investigations within the International Solar-Terrestrial Physics Program.

The contract would provide for delivery of the Wind laboratory to Cape Canaveral Air Force Station, Florida, for launch readiness on December 30, 1992, and delivery of the Polar laboratory to Vandenberg Air Force Base, California, for launch readiness on June 30, 1993. Each laboratory would be launched on a government-furnished expendable launch vehicle. (NASA News, Nov 23/88)

November 26: A Soyuz TM-7 spacecraft, carrying a joint Soviet-French crew, was successfully launched from the Baikonur Cosmodrome in Kazakhstan, for a month-long tour in the Mir space station. With French President François Mitterand in attendance, the three-man crew, which included French astronaut Jean-Loup Chretien, lifted off from Baikonur on the second French-Soviet crew-tended mission. Chretien had also been aboard the first joint mission in July 1983. The three-man Soyuz crew planned to join the two-man crew, aboard Mir since January, to perform a series of scientific experiments. Additionally, Chretien was scheduled to become the first West European to perform a spacewalk. (FBIS-Sov-88-228; SSR 1988 104A; *NY Times,* Nov 27/88; *W Post,* Nov 27/88; *CSM,* Nov 28/88)

• After 23 years in solar orbit, Pioneer 6, the oldest operating spacecraft, returned to the vicinity of its launch. The craft passed within 1.16 million miles of the Earth, roughly five times the distance between the Earth and its Moon. This distance was the closest Pioneer 6 had ever approached Earth since launch from the Kennedy Space Center (KSC), Florida in 1965. With only five lunar distances separating them, Earth's gravitational pull was strong enough to haul the craft toward its own orbit, thereby increasing time in orbit for Pioneer 6 by six days (from 311 to 317 days) and 6 million miles.

The spacecraft was designed to study the Sun's atmosphere, the heliosphere. After 23 years, two of Pioneer 6's original instruments were still transmitting information on the turbulence of the solar wind to scientists at NASA's Ames Research Center, Mountain View, California. Three sister probes followed Pioneer 6 into solar orbit. During the period from 1966 to 1970, Pioneers 7, 8, and 9 were launched into coordinated orbits. With Pioneer 6, this spacecraft team formed a network of solar weather stations monitoring the heliosphere, triangulating information back to Earth from the far side of the Sun, and increasing understanding of solar effects on Earth's magnetic field. (SSR 1965 105A; ARC Release 88-81)

November 30: Thomas L. Moser was named Deputy Associate Administrator for Space Station by the head of the Space Station Office, James B. Odom. Moser had been serving as Director of the Space Station Freedom program in Reston, Virginia, since October 1986. (NASA Release 88-163)

During November: NASA launched two heavy-lift balloons from Australia in November to continue scientific studies of Supernova 1987a. A 28.4 million cubic foot balloon was scheduled to carry a scientific experiment, weighing 1,100 pounds, to measure gamma-ray and hard x-ray emissions. A second, 29.47 million cubic foot balloon also was scheduled to carry a scientific experiment, weighing 3,700 pounds, that would measure gamma-ray emissions, but over a different energy band. These scientific missions were part of the overall NASA Balloon Program managed by the Wallops Flight Facility, Wallops Island, Virginia. (NASA Release 88-154; MSFC Release 88-160)

NASA deployed the C-141 Kuiper Airborne Observatory (KAO), to observe Supernova 1987a. KAO completed a five-week deployment to New Zealand to conduct high altitude observations of charged particle and heavy metals emissions produced in the explosion of the giant blue star, first detected in February 1987. The November 1988 mission found nickel, argon, and iron exploding outward at 868 miles per second. Previous Kuiper observations of the expanding ejected cloud had greatly contributed to understanding how the explosion proceeds. KAO Supernova 1987a research was being conducted by the Astrophysics Branch, Space Science Division, Ames Research Center, Mountain View, California. (NASA Release 88-168; ARC Release 88-89).

December

December 2–6: The Space Shuttle Atlantis was launched from Kennedy Space Center (KSC), Florida, on mission STS-27, a dedicated Department of Defense (DoD) mission. After a 24-hour launch delay because of strong winds at KSC, Atlantis lifted off on a four-day mission to deliver into orbit a classified DoD payload, reported by various media sources as being a Lacrosse radar reconnaissance satellite. Crew members for STS-27 were Robert L. Gibson, Commander; Guy S. Gardner, Pilot; and Richard M. Mullane, Jerry L. Ross and William M. Shepherd, Mission Specialists. Atlantis landed on December 6 at Edwards Air Force Base, California. (NASA Release 88-171; *AP,* Dec 2/88; *UPI,* Dec 2/88; *NY Times,* Dec 7/88; *W Post,* Dec 7/88; *WSJ,* Dec 7/88; *W Times,* Dec 7/88; *B Sun,* Dec 7/88; *P Inq,* Dec 7/88)

December 5: The Pioneer Venus orbiter spacecraft marked 10 years of orbital observations of the cloud-draped planet. Ceremonies were held at NASA's Ames Research Center, Mountain View, California, to celebrate Pioneer's accomplishments. December 9 would also mark 10 years since a companion spacecraft, Pioneer Venus 2, had entered the atmosphere of Venus along with four hard-impact probes it had released. Observations by Pioneer revealed a great deal about the atmospheric content of Venus, plate tectonics, volcanic activity, and the magnetic field. Pioneer had also observed five comets, including Halley. Pioneer, whose orbit was deteriorating, was scheduled to cease operating in August 1992 and be destroyed upon entering the lower atmosphere of Venus. (NASA Release 88-165)

December 8: NASA and the Missiles Division of LTV Missiles and Electronic Group, Dallas, Texas, announced the signing of an agreement that would grant the firm exclusive rights to commercially produce and market the Scout launch vehicle. NASA considered the negotiation of the agreement as another important advance in establishing a strong U.S. commercial launch vehicle industry through the privatization of Government expendable rocket programs. (NASA Release 88-167)

December 11: The European Space Agency (ESA) successfully launched its first Ariane 4 booster from the Korou launch center in French Guiana. The Ariane 4, the latest and most powerful ESA-built booster, carried two European telecommunications satellites. The launch had been delayed by 24 hours because of a faulty sensor on the rocket's third stage. (FBIS-Weu-88-238, Dec 11/88; SSR 1988 109C-D; *NY Times,* Dec 12/88)

December 21: Two Soviet cosmonauts and a French astronaut returned to Earth aboard a Soyuz TM-6 capsule, following the completion by the Soviets of a record breaking year-long endurance mission aboard the Mir space station. Cosmonauts Vladimir Titov and Musa Manarov, launched into space on December 21, 1987, and French astronaut Jean-Loup Chretien, successfully landed their capsule in Kazakhstan. Titov and Manarov were to undergo extensive medical tests to determine the effects of prolonged weightlessness. (FBIS-Sov-88-245, Dec 21/88; *NY Times,* Dec 22/88; *W Post,* Dec 22/88; *WSJ,* Dec 22/88; *CSM,* Dec 22/88; *B Sun,* Dec 22/88; *P Inq,* Dec 22/88)

December 29: E. Ray Tanner was appointed Director of the Space Station Freedom Program Office, effective January 3, 1989. Tanner had been serving as Manager of the Space Station Projects Office, Marshall Space Flight Center, Huntsville, Alabama, since August 1988. Prior to that he was Deputy Director for Space Systems in the Science and Engineering Directorate. (NASA Release 88-175; MSFC Release 88-178)

• Margaret G. Finarelli was appointed Deputy Associate Administrator for External Relations, effective December 12. Finarelli had been serving as Director of the Policy Division in the Office of Space Station. (NASA Release 88-164)

January

January 3: A NASA ER-2 high altitude aircraft flew the first of a dozen scientific missions over the arctic circle to document the depletion of the Earth's atmospheric ozone layer around the North Pole. The project, a joint U.S.-European effort, involved over 150 scientists engaged in research on ozone depletion caused by chlorofluorocarbons (CFCs) and other man-made chemicals. (*AP,* Jan 3/89; *USA Today,* Jan 4/89)

January 5: Dr. C. Howard Robins, Jr. was named NASA Associate Administrator for Management, effective January 9. His primary responsibility was to help strengthen NASA's core organizational and infrastructural capabilities. Robins, a 30-year NASA veteran, would replace Manuel Peralta, who left NASA to become president of the American National Standards Institute. (NASA Release 89-1, Jan 5/89; *W Times,* Jan 11/89)

January 11: A group of U.S. and Canadian astronomers announced that they had discovered evidence of vigorous activity in a white-dwarf star previously thought to be a stellar "corpse". This new and unexpected behavior may offer astronomers new insight into how stars are born, evolve, and ultimately die. The white-dwarf star, catalogued as 0950+139, lies at the center of the faint nebula called EGB-6 and is located about 1500 light years from Earth in the direction of the constellation Leo. The findings were presented at the 173rd meeting of the American Astronomical Society in Boston, Massachusetts. (NASA Release 89-3, Jan 11/89)

• NASA issued a request for proposals to design and build a ground-based radar that would quantify and characterize debris orbiting between 180 to 360 miles above Earth. The radar would have the capability of detecting debris as small as one centimeter in diameter, contrasted with the 10-centimeter capability of current radar systems. The data gathered by the orbital debris radar would be used in designing the permanently crew-tended Space Station Freedom, which would be built to withstand as much orbital debris damage as possible. (NASA Release 89-8, Jan 19/89; *W Times,* Jan 23/89; *NY Times,* Jan 24/89; *P Inq,* Jan 29/89)

January 20: Morton Thiokol conducted a full-scale test firing of a Space Shuttle solid rocket motor (SRM), Qualification Motor 8 (QM-8), under conditions simulating a cold-weather launch. The QM-8 was the sixth and final test of the redesigned Space Shuttle SRMs. A major objective of the test was to determine the performance of the SRM's redesigned field joints under cold weather conditions. (*LA Times*, Jan 21/89, *P Inq*, Jan 21/89)

January 21: The National Weather Service's Geostationary Operational Environmental Satellite-6 (GOES-6) ceased to function, creating a temporary gap in U.S. satellite weather coverage. The 6-year-old GOES-6, which had surpassed its expected operational lifetime by a year, was to be temporarily replaced by a similar satellite already in orbit. The GOES-7 satellite was to be shifted eastward to cover the entire United States until a permanent replacement, GOES-I, was launched in 1990. (*UPI*, Jan 23/89; *W Post*, Jan 23/89; *W Times*, Jan 23/89; *NY Times*, Jan 24/89)

January 27: A European Space Agency Ariane 2 launch vehicle was successfully launched from the Kourou launch center in French Guiana. The Ariane 2 carried a two-ton Intelsat-V telecommunications satellite. (SSR 1989 006A-B; *W Times*, Jan 27/89)

January 29: The Soviet Union's Phobos 2 Mars observer spacecraft entered into Martian orbit after firing its braking engine. In addition to various instruments for remote observation of Mars, Phobos 2 carried a lander module to be deposited on Mar's moon Phobos in early April. Phobos 1 became inoperative in August 1988 after a flight command error disoriented the spacecraft and misdirected its solar power arrays. The schedule for Photos 2 was moved forward by several days because of electrical problems that might endanger the mission. (FBIS-SOV-89-018, Jan 30/89; *NY Times*, Jan 30/89; *LA Times*, Jan 30/88; *P Inq*, Jan 30/89)

During January: Using data from the orbiting Infrared Astronomical Satellite (IRAS), astronomers at the University of Michigan at Ann Arbor discovered several dim galaxies in an area of the universe previously believed to be devoid of large celestial objects. The discovery of seven dim galaxies in the "Bootes Void" challenges accepted theories of the structure of the universe, which is believed to consist of filaments or "bubbles" of matter surrounding less dense areas of relatively empty space. The Michigan astronomers said they planned to continue studying IRAS's infrared data in hopes of finding more dim galaxies. (*W Post*, Jan 16/89)

February

February 8: Astronomers from the University of California at Berkeley announced the possible discovery of a newborn pulsar at the heart of stellar debris from Supernova 1987a. A pulsar is an extremely dense sphere of neutrons less than a few miles in diameter but with a magnetic field trillions of times more powerful than the Earth's. The object at the heart of Supernova 1987a appeared to be spinning at a rate of 2,000 rotations per second. The observations were made by an international team of scientists working at the Cerro Tololo Inter-American observatory in Chile. The frequency of the light emitted by the object was found to vary in a regular pattern, suggesting that it was part of a binary system, with a smaller object rotating around a larger twin. The discovery was announced in a circular sent to members of the International Astronomical Union. (*UPI,* Feb 8/89; *NY Times,* Feb 9/89; *B Sun,* Feb 10/89)

February 10: The Soviet Union launched a cluster of six Cosmos satellites aboard a Cyclone launch vehicle, including an Antarctic observer spacecraft that was placed into near-polar orbit, the Tass News Agency reported. The Cosmos 2000 satellite was described as a photo-reconnaissance research satellite that would photograph unexplored regions of central Antarctica. (SSR 1989A-G; *UPI,* Feb 11/89)

February 14: The Air Force successfully launched the first of its new Delta 2 launch vehicles, carrying a $65 million NAVSTAR satellite into a preliminary elliptical orbit. The NAVSTAR, the first operational satellite of the Global Positioning System (GPS), subsequently boosted itself into an 11,000 mile circular orbit. The 21-satellite GPS network was expected to provide very precise positioning data for U.S. and North Atlantic Treaty Organization (NATO) troops and vehicles throughout the world. The satellites would also broadcast precise positioning data to civilian users. (SSR 1989 013A; *UPI,* Feb 14/89; *W Post,* Feb 15/89; *NY Times,* Feb 15/89; *LA Times,* Feb 15/89; *W Times,* Feb 15/89; *USA Today,* Feb 15/89; *P Inq,* Feb 15/89; *W Times,* Feb 17/89)

The U.S. Government released results of a six-month interagency study on orbital debris. The study, cochaired by NASA and the Department of Defense, cited satellite and rocket body fragmentation as the principal source of orbital debris and concluded that, left unchecked, the growth of debris could threaten the safe and reliable operation of crew-tended and untended spacecraft in the next century. (NASA Release 89-20, Feb 17/89)

February 21: In civil proceedings against Rockwell International Corporation in a Los Angeles Federal court, prosecutors recommended that a $5 million fine be levied against the company, after pleading guilty to overbilling the Government on a satellite navigation contract. Rockwell had entered a conditional guilty plea for overbilling the Air Force by nearly $450,000 on contracts for the $1.2 billion NAVSTAR satellite system. (*UPI,* Feb 21/89; *LA Times,* Feb 22/89)

• The Tass News Agency reported that the Soviet Phobos 2 spacecraft had begun to transmit high quality closeup images of the Mars moon Phobos. Phobos 2 was expected to dispatch two small robot landers to the surface of Phobos. (FBIS-Sov-89-034; *LA Times,* Feb 23/89; *W Post,* Feb 27/89)

February 27: Morton Thiokol announced that it would divide its operations into two companies: Morton International, Inc. and Thiokol Corporation. Morton International would include the company's salt, specialty chemicals, and automotive airbag operations, whereas Thiokol would include aerospace operations. According to Chief Executive Officer Charles Locke, the decision to split the operating units was made "so that each can adopt strategies and pursue objectives appropriate to its specific businesses." The restructuring was expected to be completed by July 1, pending stockholder approval and Internal Revenue Service (IRS) certification of the spinoff as tax-free. (*WSJ,* Feb 28/89; *W Post,* Feb 28/89; *USA Today,* Feb 28/89)

March

March 6: After several delays, an ESA Ariane 4 launch vehicle was successfully launched from the Kourou launch center in French Guiana. The 195-foot Ariane 4 carried Japan's first commercial communications satellite and a European weather station into low Earth orbits, from where the two satellites would later boost themselves into 22,300-mile geosynchronous orbits. This was the third launch of the new Ariane 4 launch vehicle, one of the most powerful boosters available for commercial satellite launches. (SSR 1989 020A-D; *UPI*, Mar 6/89)

March 13–19: The Space Shuttle Discovery was successfully launched from the Kennedy Space Center (KSC) on mission STS-29, the main objective to deploy the final Tracking and Data Relay Satellite (TDRS-4). Discovery carried a crew of five on a 5-day mission. Six hours into the mission, the 2.5 ton $100 million TDRS-4 was successfully deployed into low-Earth orbit where it subsequently boosted itself into a 22,300 mile geosynchronous orbit. Discovery also carried biological experiments that to conduct studies of the embryonic development of chickens and bone healing in rats in a microgravity environment. The STS-28 crew would also photograph and film environmentally damaged areas throughout the world for later scientific analysis. Discovery's launch had been delayed a month because of cracks found in the main engine turbopumps. Replacement occurred on the launch pad. During the mission, a hydrogen tank supplying the Shuttle's fuel cells was temporarily shut down because of erratic pressure readings. The problem, which caused a temporary electricity shortage aboard the Shuttle, was resolved the following day, allowing Discovery to stay in orbit for five days as planned. Discovery successfully landed at Edwards Air Force Base, California, on March 19. (*NASA MOR/Flight Operation Report, M-989-88-29;* SSR 1989 021A; *AP,* Mar 13–19/89; *UPI,* Mar 13–19/89; *NY Times,* Mar 14–20/89; *WSJ,* Mar 14–20/89; *W Post,* Mar 14–20/89; *W Times,* Mar 14–20/89; *USA Today,* Mar 14–20/89; *P Inq,* Mar 14–20/89; *B Sun,* Mar 14–20/89)

March 14: NASA Administrator Dr. James C. Fletcher and the Ambassador of Japan to the United States H.E. Nobuo Matsunaga signed a Memorandum of Understanding between NASA and the government of Japan on cooperation in the detailed design, development, operation, and utilization of the Space Station Freedom. The agreement was signed at NASA Headquarters, Washington, D.C. Under the agreement, Japan would provide the Japanese Experiment Module (JEM) to the Freedom program. The JEM, to be permanently attached to the Space Station base, would consist of a pressurized laboratory module, at least two experiment logistics modules, and an exposed facility. (NASA Release 89-32, Mar 14/89)

March 21: NASA Administrator Dr. James C. Fletcher announced his resignation, effective April 8. Dr. Fletcher had served as NASA Administrator for a total of nine years, from April 1971 to May 1977, and again since May 1986, following the Challenger accident. Dr. Fletcher, who reluctantly agreed to return to the Agency at the request of President Ronald Reagan, supervised the adoption of a new management system that put greater emphasis on quality control and safety for the Shuttle program. During his second term, Dr. Fletcher had oversight of the redesign of the Space Shuttle solid rocket boosters and successfully lobbied Congress for continuation of the Space Station program. In his resignation announcement, Dr. Fletcher said he now felt that he could "safely place the leadership of NASA in another's hands." NASA Deputy Administrator Dale Myers served as Acting Administrator until a successor was named by President George Bush. (NASA Release 89-36, Mar 21/89; *AP,* Mar 21/89; *UPI,* Mar 21/89; *NY Times,* Mar 22/89; *WSJ,* Mar 22/89; *USA Today,* Mar 22/89; *B Sun,* Mar 22/89; *LA Times,* Mar 22/89)

• In its first underwater test firing, a Trident II missile exploded in flight four seconds after breaching the ocean surface near Cape Canaveral, Florida. Shortly after being fired from the submarine USS Tennessee, the Trident II veered off course and self destructed. Neither the submarine nor its support ships were damaged. (*NY Times,* Mar 22/89; *W Post,* Mar 22/89; *W Times,* Mar 22/89; *LA Times,* Mar 22/89)

March 23: Three-time spaceflight veteran Navy Captain Frederick Hauck, Commander of the first post-Challenger Shuttle mission, announced he would leave NASA on April 3 to join the Pentagon staff of the Chief of Naval Operations. (NASA Release 89-39, Mar 23/89)

March 24: In a major test of the Strategic Defense Initiative (SDI), a Delta Star SDI test satellite was successfully launched aboard a Delta launch vehicle at 4:51 p.m. from Cape Canaveral Air Force Station, Florida. The three-ton satellite would be used to test remote sensing equipment used to detect rocket launches and laser beam emissions from Earth in conditions that would simulate the initial stages of a nuclear conflict. The three-ton satellite carried a laser radar, seven video imaging cameras, and an infrared imager. The Delta Star was expected to carry out tests throughout the next six to nine months. (SSR 1989 026A; *NY Times,* Mar 25/89; *W Post,* Mar 25/89; *C Trib,* Mar 25/89)

March 27: Challenging earlier reports that a mysterious object at the heart of the remnants of Supernova 1987A is a rapidly spinning pulsar, scientists at Columbia University announced that the observed object was more likely to be a vibrating neutron star. The debate over the nature of the object began in February when astronomers at the University of California at Berkeley theorized that they were observing an extremely dense pulsar rotating at 2,000 rotations per second. The Berkeley announcement prompted astronomers at Columbia University to argue that the observed bursts of radiation were more likely to emanate from a less dense neutron star that is vibrating at the observed frequency. The Columbia astronomers argued that the vibration theory not only explained the recent observations but also was consistent with previous hypotheses on how neutron stars develop. (*W Post,* Mar 27/89)

March 28: In a severe setback for the Soviet Union's planetary exploration program, Soviet ground controllers lost stable radio contact with the untended Phobos 2 Mars probe, the Tass news agency reported. Phobos 2 stopped responding to ground control commands shortly after the craft was ordered to perform a delicate maneuver around the Mars moon Phobos. The maneuver was an initial step toward landing a probe on the Martian moon. During its transit toward Mars, Phobos 2 had experienced a failure of its main radio transmitter, forcing ground controllers to rely on a low-power backup transmitter for all communications with the spacecraft. A companion spacecraft, Phobos 1, was also lost in September as a result of a ground control command error. The loss of communications with Phobos 2 was described by Western experts as having occurred at the "worst possible time" because few of the main mission objectives had yet been fulfilled. (FBIS-Sov-89-059, Mar 29/89; *UPI,* Mar 28/89; *AP,* Mar 29/89; *NY Times,* Mar 29/89; *W Post,* Mar 29/89; *W Times,* Mar 29/89; *P Inq,* Mar 29/89)

March 29: The first commercial launch of a licensed private rocket was successfully carried out at White Sands Missile Range, New Mexico. A two-stage Starfire suborbital rocket built by Space Services, Inc. carried a canister of experiments on a 198-mile ballistic trajectory. The experiment canister, which carried several microgravity materials processing experiments, was retrieved intact at the conclusion of the 15-minute suborbital flight. Space Systems, Inc. had previously launched a test rocket carrying a dummy payload in 1982. (*NY Times,* Mar 30/89; *W Post,* Mar 30/89; *WSJ,* Mar 30/89; *W Times,* Mar 30/89; *B Sun,* Mar 30/89; *P Inq,* Mar 30/89; *C Trib,* Mar 30/89)

March 29: NASA scientists reported that a group of eight newly fertilized chicken embryos carried aboard the Space Shuttle Discovery on mission STS-28 had failed to hatch upon returning to Earth, raising questions about the viability of animal reproduction in zero gravity. A group of 16 eggs fertilized nine days before launch were all successfully hatched. Of a third group of eight eggs scheduled to hatch Saturday, only a few embryos remained viable. Discovery astronaut James Bagian speculated that the results may indicate that gravity plays a previously unknown critical role during the process of cell differentiation. Bagian suggested that, should the experiment demonstrate a relationship between gravity and embryogenesis, the results would have "great implications" for future long-duration space missions. (*UPI,* Mar 29/89; *W Post,* Mar 30/89; *P Inq,* Mar 30/89; *NY Times,* Mar 31/89)

March 30: NASA Administrator Dr. James C. Fletcher named Dr. Robert Rosen as Acting Associate Administrator for the Office of Aeronautics and Space Technology (OAST), effective April 2. Rosen had served as Deputy Associate Administrator, OAST, since March 1986. (NASA Release 89-41, Mar 30/89)

April

April 2: An ESA Ariane 2 launch vehicle was successfully launched from the Kourou launch facility in French Guiana. The Ariane 2 carried the TELE-X Nordic communications satellite into a preliminary elliptical orbit, from where it subsequently boosted itself into a 22,300-mile geosynchronous orbit. (FBIS-WEU-89-062, Apr 3/89; SSR 1989 027A-B; *UPI,* Apr 1/89)

April 11: The National Research Council (NRC) recommended that NASA abandon plans to participate in development of a commercial orbiting space laboratory that had been promoted by the Reagan administration and some members of Congress. In recommending against the development of the Industrial Space Facility, the NRC argued that experiments foreseen for the space laboratory could be performed aboard the Space Shuttle and on Shuttle-based facilities during the period leading to the completion of Space Station Freedom. (*NY Times,* Apr 12/89; *W Post,* Apr 12/89)

April 12: The Soviet Union announced that it would suspend occupation of the Mir Space Station for at least three months following the return to Earth of its current crew in late April. The reason given by the Soviets for the abandonment of the station, which had been continuously occupied since February 1987, was a delay in the manufacture of additional pressurized modules to be attached to the facility. Western experts were interpreting the temporary shutdown of the station as a possible cost-cutting measure and another setback to the Soviet space program following the loss of its twin Mars probes during the last several months. *Aviation Week & Space Technology* magazine was also reporting that degradation of the station's solar power arrays was causing an electrical shortage. (*NY Times,* Apr 13/89; *W Post,* Apr 13/89, 15/89; *LA Times,* Apr 12/89; *UPI,* Apr 14/89; *W Times,* Apr 14/89; *P Inq,* Apr 19/89)

April 13: President George Bush announced the selection of Rear Admiral Richard H. Truly, a former astronaut and head of the Space Shuttle program, to succeed Dr. James C. Fletcher as NASA Administrator. The White House also announced that J.R. Thompson, Director of the Marshall Space Flight Center (MSFC), Huntsville, Alabama, would be appointed NASA Deputy Administrator. Truly, a veteran of two Shuttle flights, was named to head the Shuttle program by Administrator Fletcher following the 1986 Challenger accident. Truly was credited with helping to restore safety and reliability to Shuttle-related activities. (FBIS-Sov-89-070, Apr 13/89; *AP,* Apr 12/89; *UPI,* Apr 12/89; *NY Times,* Apr 13/89; *W Post,* Apr 13/89; *WSJ,* Apr 13/89; *USA Today,* Apr 13/89; *W Times,* Apr 13/89)

• Dale D. Myers, the acting NASA Administrator, announced his plans to resign effective May 13. Myers had served as the NASA Deputy Administrator from October 6, 1986, when he was recruited to return to NASA by President Ronald Reagan. During his tenure, Myers was instrumental in guiding NASA through the period of recovery following the Challenger accident of January 28, 1986. (NASA Release 89-49, Apr 13/89)

April 16: The Soviet weekly magazine *Ogonyok* revealed additional details on the circumstances surrounding a mysterious launch pad explosion of a Soviet rocket in 1960 that killed dozens of workers and the commander of the Soviet Union's rocket forces. Ogonyok attributed the accident, which took place at a secret launch facility at Tura-Tam, near the Aral Sea, to a flouting of safety rules during a crash program to catch up with the United States in the development of Intercontinental Ballistic Missiles (ICBMs). The head of the Soviet Rocket Forces, Chief Artillery Marshal Mitrofan I. Nedelin, was killed along with dozens of workers and soldiers who were situated near the pad when the rocket exploded. (*AP,* Apr 17/89; *B Sun,* Apr 17/89; *W Times,* Apr 17/89; *C Trib,* Apr 17/89)

April 19: Scientists at NASA's solar system exploration division announced that the Earth had experienced a "close call" in March with a newly discovered half-mile-wide asteroid. The asteroid, named 1989FC, passed within 500,000 miles of Earth—about twice the distance to the Moon—on March 23. The asteroid was discovered by Henry Holt, an amateur astronomer working on a NASA asteroid hunting project. NASA scientists estimated that if the asteroid were ever to collide with the Earth, the collision would release energy equivalent to 20,000 times that of the atomic bomb dropped on Hiroshima and would leave a crater 10 miles wide and up to a mile deep. (NASA Release 89-52, Apr 19/89; UPI, Apr 19/89; AP, Apr 20/89; NY Times, Apr 20/89; *USA Today,* Apr 20/89; *LA Times,* Apr 20/89; *B Sun,* Apr 20/89)

April 19: NASA rolled out a full-scale mockup of the proposed Shuttle-C, a new heavy lift launch vehicle based on Space Shuttle technology. The 115-foot mockup, constructed in a Hangar at the Marshall Space Flight Center, Huntsville, Alabama, was put on display in an effort to gain support for funding of the $1.5 billion Shuttle-C development program in the fiscal year 1991 Federal budget. The Shuttle-C would be an untended launch vehicle that would fly "piggyback" on the standard Shuttle SRM and external tank assembly and would have a greater payload capacity than the current Shuttles. Shuttle-C could substitute for the crew-tended Shuttles in the launching of components for Space Station Freedom. (*AP,* Apr 18/89; *LA Times,* Apr 19;/89)

April 20: President George Bush reestablished the National Space Council, an executive space policy advisory group to be chaired by the Vice President.

The Space Council was originally created by President John F. Kennedy and disbanded in 1973 by President Richard Nixon. President Bush restored the Space Council as an official forum for setting long-range policy for United States civil, military, and commercial space efforts. (WH Release, Apr 20/89; *UPI*, May 12/89)

• NASA announced that Martin Marietta Corporation of Bethesda, Maryland, had won a $297 million contract to develop the Flight Telerobotic Servicer (FTS), a space robot that would help construct and maintain the Space Station Freedom. Development of the robot was expected to take nine years and would include flight testing of a prototype aboard the Space Shuttle. Once deployed during an early Space Station construction mission, the mobile robot would be controlled remotely by astronauts from within the station, thereby minimizing the amount of "spacewalk" activity that astronauts would need to perform. (*W Times*, Apr 21/89; *B Sun*, Apr 21/89)

April 22: NASA announced that it had awarded a $1.1 billion contract to Lockheed Corporation of Calabasas, California and GenCorp, Inc. of Fairlawn, Ohio, for the development of the Space Shuttle advanced solid rocket motor (ASRM). The ASRM, which would replace the current Shuttle SRMs built by Morton Thiokol, would be a next-generation solid fuel propulsion system that would use an elongated single rocket casing instead of the current segmented case design. According to NASA, the new design would improve Shuttle safety and performance, allowing an additional 12,000-pound payload capacity aboard the orbiters. Lockheed and GenCorp would eventually replace Morton Thiokol as prime contractors for Shuttle solid rocket propulsion. A faulty field joint on one of Thiokol's segmented boosters was blamed for the catastrophic explosion of the Space Shuttle Challenger in 1986. Under the terms of the seven-year contract, Lockheed's Missile System division and GenCorp's Aerojet Space Booster Company would design, develop, and test the new generation of motors, and produce enough for six Shuttle flights beginning as early as 1994. The contract also had an option, valued at approximately $1 billion, for NASA to purchase ASRMs for 40 additional Shuttle flights and eight test firings. (NASA Release 89-57, Apr 21/89; *UPI*, Apr 22/89; *NY Times*, Apr 22/89; *W Post*, Apr 22/89; *LA Times*, Apr 22/89; *WSJ*, Apr 24/89)

April 24: Six top-level NASA managers announced their resignations in order to avoid new conflict-of-interest restrictions on contract related work by former Federal employees. The new restrictions, effective May 16, 1989, would bar officials who have worked on procurement from assuming any private sector jobs involving the contracts for two years. Some of the resigning managers cited the failure of Congress to pass a pay raise for senior executive service (SES) personnel as a reason for leaving. The six NASA officials who announced their resignations were Noel Hinners, Associate Deputy Administrator; James Odom, Director of the Space Station program; Jon McBride, an astronaut and chief of congressional relations; John Thomas, head of solid rocket booster redesign at the Marshall Space Flight Center (MSFC); Bill Sneed, MSFC policy chief; and James Downey, irector of MSFC's payload project office. (NASA Releases 89-60, 89-61, 89-62, Apr 24/89; *AP*, Apr 24/89; *UPI*, Apr 24/89; *WSJ*, Apr 25/89; *W Post*, Apr 25/89; *W Times*, Apr 25/89)

April 25: The Department of Defense (DoD) announced it would phase out its participation in the National Aerospace Plane program, reducing funding for the project by $200 million in fiscal year 1990 and obligating its remaining $100 million contribution to NASA. The National Aerospace Plane was proposed by President Reagan as a hypersonic vehicle that would be capable of taking off from conventional runways and subsequently boosting itself into low-Earth orbit. Total project development costs were estimated at $3.5 billion, of which 80 percent would come from the DoD and 20 percent would be provided by NASA. DoD and NASA had already invested $850 million to develop an X-30 experimental prototype vehicle. (*UPI*, Apr 25/89)

April 27: Three Soviet cosmonauts who had been occupying the Mir Space Station returned to Earth aboard a Soyuz TM-7 capsule. The three cosmonauts landed safely in Soviet Central Asia near the city of Dzheskazgan. The departure of the Mir crew marked the first time that the station has been left untended since crew rotations began in February 1987. Soviet space program officials announced that two new experiment modules would be attached to the station in anticipation of the arrival in August of the next crew. (FBIS-Sov-89-080, Apr 27/89; *UPI*, Apr 27/89; *P Inq*, Apr 28/89; *B Sun*, Apr 28/89)

April 28: A launch attempt of the Space Shuttle Atlantis was scrubbed 31 seconds before scheduled liftoff because of a power surge on a pump that recirculates liquid hydrogen fuel for one of the Shuttle main engines. Shuttle technicians were working to replace the pump and a leaky fuel line that was discovered after the scrub. A new launch attempt was expected to be made as early as Thursday. (*UPI*, Apr 29–30/89; *AP*, Apr 30/89; *W Post*, May 1/89; *NY Times*, May 1/89; *USA Today*, May 1/89; *B Sun*, May 1/89; *P Inq*, May 1/89)

May

May 4–8: The Space Shuttle Atlantis was successfully launched from the Kennedy Space Center, Florida, on mission STS-30, whose main objective was to deploy the $550 million Magellan Venus radar mapper. Magellan, the first United States interplanetary mission to be launched in 11 years, was deployed from the cargo bay of Atlantis 6 hours after launch and was subsequently propelled toward the Sun and Venus by a solid fuel inertial upper stage (IUS). Following the IUS firing, ground controllers confirmed that the spacecraft was operating normally and was on a proper course toward Venus. Upon arriving at Venus in August 1990, the 7,700-pound Magellan would enter Venusian orbit for several months of extensive radar mapping of the planet's cloud shrouded surface. Atlantis landed at Edwards Air Force Base, California, on May 8. (NASA MOR, C-600-89-30, E-199-89-30; NASA PFOR, M-989-89-30; SSR 1989 033A-B; *UPI,* May 5/89; *NY Times,* May 5/89; *W Post,* May 5/89; *WSJ,* May 5/89; *P Inq,* May 5/89; *B Sun,* May 5/89; *W Times,* May 5/89; *USA Today,* May 5/89)

May 9: The Department of Defense announced that it would terminate funding for development of the V-22 Osprey tilt-rotor aircraft beginning in FY 1990. The V-22 Osprey had been under development by Boeing Helicopters and Bell Helicopter Textron, Inc. for the U.S. Marine Corps. (*P Inq,* May 10/89)

May 10: An Air Force Titan 34D launch vehicle was successfully launched from Cape Canaveral Air Force Station, Florida, carrying a secret military payload into orbit. The 16-story, $65 million Titan was believed to be carrying a pair of Defense Satellite Communications System (DSCS) satellites. (SSR 1989 035A-C; *W Times,* May 11/89; *C Trib,* May 11/89; *B Sun,* May 11/89; *P Inq,* May 11/89)

May 10: President George Bush chose the name "Endeavour" for the new Space Shuttle orbiter being built to replace the orbiter Challenger. The name "Endeavour" resulted from a nationwide orbiter-naming competition supported by educational projects created by student teams in elementary and secondary schools. Endeavour was the name of the first ship commanded by James Cook, a British explorer, navigator, and astronomer. Cook commanded the Endeavour on its maiden voyage to the South Pacific on an astronomical and mapping mission. The new orbiter, designated OV-105, was scheduled to be completed in 1991 and to fly its maiden voyage in March 1992. (NASA Release 89-70, May 10/89)

May 11: An unarmed Air Force Midgetman missile went off course and was destroyed shortly after launch from Vandenberg Air Force Base. The missile began to tumble 70 seconds into the flight during its second stage burn, causing the range safety officer to transmit a self-destruct command. The Midgetman, which was undergoing its first flight test, had been programmed to deliver a dummy warhead 4,600 miles over the Pacific Ocean to the Kwajalen Atoll test range. Despite the failure, the test was declared a partial success by the Air Force. (*NY Times,* May 12/89; *WSJ,* May 12/89; *USA Today,* May 12/89; *W Times,* May 12/89)

May 12: Thomas L. Moser, acting Associate Administrator for the Space Station Freedom program, announced he was leaving NASA effective May 13. Moser, appointed Deputy Associate Administrator for the Freedom program in December, had been serving as the acting Associate Administrator since April 30. (NASA Release 89-72, May 12/89)

May 15: The Federal Aviation Administration (FAA) reported that a NASA T-38 jet trainer aircraft being flown by a Space Shuttle astronaut had experienced a near collision with a Pan American World Airways widebody jet about 30 miles West of Washington, DC. The pilots of the Airbus A-310 jetliner reported the two planes had come within 500 feet of each other while on perpendicular flight paths. The T-38 was flown by Navy Captain David M. Walker, Commander of the recently completed Space Shuttle mission STS-28. The near collision occurred when Walker's aircraft strayed below its assigned flight elevation. (*AP,* May 17/89; *UPI,* May 17/89; *NY Times,* May 17/89; *W Post,* May 17/89; *B Sun,* May 17/89)

May 16: Acting NASA Administrator Richard H. Truly announced the appointment of Samuel W. Keller as NASA Deputy Associate Administrator, the third-ranking position in the Agency. Keller had been Deputy Associate Administrator for Space Science and Applications since December 1977. (NASA Release 89-75, May 16/89; *LA Times,* May 17/89)

May 18: Acting NASA Administrator Richard H. Truly announced the appointment of former Space Shuttle astronaut William B. Lenoir as Associate Administrator for Space Station. Lenoir replaced James B. Odom, who was retiring. Truly also announced the appointment of Richard Kohrs as Director, Space Station Freedom. (NASA Release 88-77, May 18/89; *UPI,* May 18/89; *AP,* May 19/89; *NY Times,* May 19/89; *W Post,* May 19/89; *WSJ,* May 19/89; *LA Times,* May 19/89; *W Times,* May 23/89)

May 24: The Soviet Union successfully launched a COSMOS 2021 satellite. (*AP,* May 28/89; *UPI,* May 28/89)

May 31: The Soviet Union successfully launched a Proton launch vehicle carrying three satellites from the Baikonur launch facility in Kazakhstan. The Proton was carrying the Cosmos 2,022, 2,023 and 2,024 satellites. The Tass News Agency reported that the third stage of the booster failed to burn up completely in the atmosphere and that unburned fragments may have fallen to Earth in the U.S.-Canadian border region near International Falls, Minnesota. (SSR 1989 039A-H; *UPI,* June 2/89; AP, June 2/89; *LA Times,* June 3/89)

During May: Astronomers at Washington University, St. Louis, Missouri, published results of a study which confirmed that Pluto is a planet and not an asteroid or a wayward natural satellite. William McKinnon, Associate Professor of Planetary Science at Washington University, asserted that his analysis of telescopic data and computer models showed that Pluto was formed independently in the outer solar system as a planet. (*LA Times,* May 29/89)

June

June 1: At the request of the National Space Council, President George Bush restored funding for continued operation of two Landsat satellites and the launching of a third. The National Oceanic and Atmospheric Administration had earlier decided to deactivate the Landsats because of a shortfall of $5 million in fiscal year 1989 funding for the program. (*WSJ*, Jun 2/89)

June 5: An ESA Ariane 4 launch vehicle was successfully launched from the Kourou launch facility in French Guiana, carrying two communications satellites. The 192-foot Ariane 4 was carrying a Superbird-A communications satellite owned by Japan and a West German DFS Kopernikus-1 television relay satellite. This was the 27th successful launch by Arianespace and the fourth successful launch of the new Ariane 4 launch vehicle. (FBIS-Weu-89-107; SSR 1989 041A-D; *UPI*, Jun 5/89; *P Inq*, Jun 6/89)

June 8: This date marked the thirtieth anniversary of the first hypersonic flight of the X-15 experimental rocket plane. North American Aviation pilot A. Scott Crossfield was at the controls for the first flight. The X-15 research aircraft flew a total of 199 flights in what is widely regarded as one of the most successful aeronautical research programs ever conducted. (NASA Release 89-82, May 30/89)

June 9: Three-time space flight veteran George "Pinky" Nelson announced he would leave NASA on June 30 to accept academic and administrative positions at the University of Washington, Seattle. (NASA Release 89-89, Jun 9/89)

June 10: After five cancelled launch attempts, an Air Force Delta 2 launch vehicle was successfully launched from Cape Canaveral Air Force Station in Florida, carrying a NAVSTAR navigation satellite. The NAVSTAR was the second of a planned network of satellites known as the Global Positioning System, which would allow military units to determine their position anywhere on Earth within 50 feet. (SSR 1989 043D, 044A; *W Times*, Jun 12/89)

June 14: An Air Force Titan 4 launch vehicle was successfully launched from Cape Canaveral Air Force Station, Florida, carrying a classified military satellite. This was the first launch of the Titan 4, a new generation of heavy lift launch vehicles dedicated mainly to Department of Defense missions. The $220 million Titan 4 was believed to be carrying a $180 million advanced early warning satellite. (SSR 1989 046A-E; *NY Times*, Jun 15/89; *WSJ*, Jun 15/89; *W Post*, Jun 15/89; *USA Today*, Jun 15/89; *C Trib*, Jun 15/89; *LA Times*, Jun 15/89; *W Times*, Jun 15/89)

• A team of astronomers reported that they had discovered evidence of "brown dwarfs", large celestial objects that are intermediate in size between planets and stars. The findings, presented at a meeting of the American Astronomical Society, showed evidence of at least nine brown dwarfs orbiting stars in the constellation Taurus. The observed objects were believed to be from 5 to 20 times the size of Jupiter. Evidence of brown dwarfs was expected to help account for some of the "missing" dark matter believed to exist in the universe. (*W Post*, Jun 15/89; *P Inq*, Jun 25/89)

June 17: Space Shuttle astronaut David S. Griggs was killed while off duty when a single engine propeller plane he was flying crashed near a private airstrip in Arkansas. Griggs had been performing aerobatic maneuvers in a vintage ST-6 training plane when the craft struck the ground, killing the astronaut instantly. Griggs was a mission specialist aboard an April 1985 flight of the Space Shuttle Discovery and had been designated pilot for Discovery mission STS-33 in November. (*AP*, Jun 18/89; *W Post*, Jun 18/89; *UPI*, Jun 19/89; *NY Times*, Jun 19/89; *P Inq*, Jun 19/89)

June 19: Scientists at NASA's Jet Propulsion Laboratory, Pasadena, California, announced that the Voyager 2 space probe had discovered a giant weather system on Neptune approximately 6,200 miles wide and comparable to Jupiter's Great Red Spot. Voyager 2 made the discovery while at a distance of 58.98 million miles from Neptune, currently the most distant planet from the sun. Voyager 2 was scheduled to perform an extremely close flyby of Neptune on August 24, when it would approach to within 3,000 miles of the planet's cloud tops. This would be the fourth and final planetary flyby by Voyager 2, which was launched in 1977 and had flown by Jupiter, Saturn, and Uranus. (*Voyager Bulletin: Mission Status Report No. 91*, Aug 17/89; *AP*, June 16/89; *W Post*, Jun 21/89)

• The first major private-sector launch of a satellite was delayed by several months as a result of a launch pad accident that damaged the satellite. A communications satellite owned by the government of India was seriously damaged when it was accidentally struck by a lift hook connected to a crane. The satellite was being prepared for launch at a McDonnell Douglas facility at Cape Canaveral, Florida. (*W Post*, Jun 27/89; *O Sent Star*, Jun 21/89)

June 23: A Space Shuttle main engine shut down prematurely and suffered severe damage during a full-duration test firing at the Stennis Space Center in Mississippi. NASA engineers said that a high speed liquid oxygen pump "came apart" during the test firing, causing a fire and structural damage to Rocketdyne engine No. 0212. This was the first serious malfunction of a main engine since 1985. (MSFC Release 89-125, Jun 26/89; *UPI*, Jun 23/89; *NY Times*, Jun 28/89; *W Times*, Jun 28/89)

June 26: Planetary scientists at NASA's Jet Propulsion Laboratory announced that Earth-based radar observations of Saturn's moon Titan had shown that it is not entirely covered by an ocean of liquid ethane. Radar echoes bounced from Titan between June 3 and 6 showed the planet to have some dry land surfaces. Previous observations by the Voyager 2 space probe had shown that Titan had large oceans of liquid ethane and methane. (*W Times,* Jun 27/89)

June 27: NASA director of the Space Station Freedom Program Office, Ray Tanner, announced his resignation from the space agency, citing new ethics rules on contract work by former Federal employees as the reason for his departure. Tanner was a 30-year NASA veteran. (*Wash Tech,* Jul/89)

June 28: The Japanese Science and Technology Agency announced a 10-year plan to develop an untended space plane and a series of robots capable of assembling and repairing satellites in space. The Agency also said that Japan would begin preliminary research into a crew-tended space shuttle. (*UPI,* Jun 28/89; *W Times,* Jun 30/89)

June 29: Great Britain and the Soviet Union signed an agreement to send a British astronaut to the Mir Space Station aboard a Soviet crew rotation flight in 1991. (*USA Today,* Jun 30/89; *W Times,* Jun 30/89; *LA Times,* Jun 30/89)

June 30: The National Space Council, chaired by Vice-President Dan Quayle, recommended to President George Bush that the National Aero-Space Plane program be maintained but that its activities be scaled down and focused on pure research. The new plan would cut spending for the program from $427 million to $254 million in fiscal year 1990, with funding being split evenly between the Department of Defense and NASA. (*AP,* Jul 1/89; *W Post,* Jul 1/89; *NY Times,* Jul 2/89; *CSM,* Jul 10/89)

July

July 5: Planetary scientists at NASA's Jet Propulsion Laboratory (JPL) confirmed that the Voyager 2 space probe had discovered a third moon orbiting Neptune. The moon, temporarily designated 1989 N1, was estimated to be between 125 and 400 miles in diameter and to be in a near-circular orbit 57,600 miles above Neptune's equator. Additionally, JPL announced that a pair of dark atmospheric bands had been discovered around Neptune's south pole. The scientists compared the 2,700 mile-wide belts to jet streams on Earth and to the belts seen around Jupiter, Saturn, and Uranus. The photograph showing the south pole bands also revealed evidence of comparable bands around Neptune's north pole. Voyager 2 was scheduled to fly to within 3,000 miles of Neptune on August 24. (NASA Release 89-110, Jul 7/89; JPL *Voyager Status Report,* Jul 7/89; *AP,* Jul 5/89; *W Post,* Jul 6/89; *UPI,* Jul 7/89; *NY Times,* Jul 8/89; *W Post,* Jul 8/89; *LA Times,* Jul 8/89; *C Trib,* Jul 8/89; *P Inq,* Jul 8/89)

July 6: NASA Administrator Richard H. Truly announced the appointment of Thomas J. Lee as Director of the Marshall Space Flight Center, Huntsville, Alabama, effective immediately. Lee succeeded James R. Thompson Jr., selected by President George Bush to be NASA Deputy Administrator. (NASA Release 89-108, Jul 6/89)

July 7: The director of NASA's Ames Research Center (ARC), William F. Ballhaus Jr., resigned, citing "inadequate compensation for senior Federal executives and vague new post-government employment regulations" as reasons for his departure. Ballhaus had been a NASA engineer and manager for 18 years and was serving his second term as ARC director. (NASA Release 89-111, Jul 7/89; *AP,* Jul 7/89)

July 10: The United States and West Germany signed an agreement to send two German astronauts and German scientific payloads on a future Space Shuttle mission. The German astronauts would fly aboard a 1992 Spacelab mission that would test materials processing techniques. Two Germans and a Dutch astronaut flew aboard a Shuttle Spacelab mission in November 1985. (NASA Release 89-113, Jul 10/89; *AP,* Nov 10/89; *UPI,* Nov 10/89; *P Inq,* Jul 12/89)

July 12: An ESA Ariane 3 launch vehicle carrying an experimental communications satellite was successfully launched from the Kourou launch facility in French Guiana. The 161-foot Ariane 3 placed the European Olympus 1 satellite in a preliminary orbit from which it would boost itself into a 22,300-mile geosynchronous orbit over the equator. (SSR 1989 053A-C; *UPI,* Jul 12/89)

• Robert O. Aller, NASA's Director of Space Operations, resigned to avoid falling under a new Government ethics regulation that restricts post-Federal employment on Government-related contracts. Aller became the ninth high level NASA official to resign in recent weeks and the third to cite publicly the new regulation as the cause for his departure. (*AP*, Jul 12/89; *C Trib*, Jul 12/89; *W Times*, Jul 13/89; *B Sun*, Jul 13/89)

July 16: Apollo 11 astronauts Neil A. Armstrong, Edwin E. Aldrin Jr., and Michael Collins participated in the first of several planned ceremonies at NASA facilities commemorating the twentieth anniversary of the first crew-assisted lunar landing on July 20, 1969. Armstrong, Aldrin, and Collins gathered at the Kennedy Space Center (KSC), Florida, with a crowd of 6,000 NASA employees and their families to commemorate the launching of their Saturn V rocket from KSC on July 16, 1969. The ceremony included a playback of the final three minutes of the Apollo 11 countdown and statements by the former astronauts. Following the ceremony, the astronauts rode in a 20-mile motorcade to Cocoa Beach, where they attended a luncheon in their honor. (*UPI*, Jul 16/89; *USA Today*, Jul 17/89; *B Sun*, Jul 17/89)

• NASA announced the selection of three contractors to research possible propulsion systems for the proposed Advance Launch System (ALS). NASA's Marshall Space Flight Center, Huntsville, Alabama, selected Aerojet General Corporation, Sacramento, California; Pratt and Whitney Division of United Technologies Corporation, West Palm Beach, Florida; and Rocketdyne Division of Rockwell International, Canoga Park, California, to research ALS designs. The systems definition contract totaled $20 million. The ALS would be a next-generation launch system capable of transporting 150,000-pound payloads into low Earth orbit. (*Def News*, Jul 17/89)

July 20: This date marked the twentieth anniversary of the first crew-assisted landing on the Moon by Apollo 11 astronauts Neil A. Armstrong and Edwin E. Aldrin. Armstrong, Aldrin, and their crewmate, Michael Collins, commemorated the lunar landing with ceremonies at the National Air and Space Museum, Washington, D.C., and the Johnson Space Center, Houston, Texas. During the Washington ceremony, attended by President George Bush and Vice President Dan Quayle, the President proposed a long range program of human-assisted space exploration, including a permanently habitable base on the Moon and a crew expedition to Mars, but did not establish a specific timetable. The President cited Space Station Freedom as an important stepping stone toward missions to the Moon and Mars. Speaking at the National Air and Space Museum, Neil Armstrong, the first astronaut to step on the lunar surface, said: "The Apollo program enjoyed a certain nobility of purpose—a program not to conquer enemies, but to conquer ignorance. A program not to exploit, but to explore. A program not to take from others, but to give to all, to give new knowledge, to enlarge the human horizon." (*UPI*, Jul 20/89; *NY Times*, Jul 21/89; *W Post*, Jul 21/89; *WSJ*, Jul 21/89; *USA Today*, Jul 21/89; *W Times*, Jul 21/89; *P Inq*, Jul 21/89; *B Sun*, Jul 21/89)

August

August 2: A report by the Office of Technology Assessment (OTA) entitled "Round Trip to Orbit: Human Space Flight Alternatives" recommended construction of a fifth Space Shuttle orbiter or a Shuttle-derived launch vehicle in preparation for the Space Station assembly missions of the late 1990s. Citing estimates of Space Shuttle reliability of 97 to 99 percent, OTA warned that there was an 88 percent likelihood that another Shuttle orbiter would be lost before Space Station construction is completed. The OTA report stated: "Buying more orbiters would increase the resiliency of the Space Shuttle system...[and] its ability to recover rapidly from loss of another orbiter or any other event that delays launches. The Shuttle orbiter fleet is likely to continue to suffer occasional attrition." (*AP,* Aug 2/89; *UPI,* Aug 2/89; *W Post,* Aug 3/89)

August 3: Planetary scientists at NASA's Jet Propulsion Laboratory in California announced the discovery of three additional moons of Neptune, bringing the total number of Neptune's known moons to six. The moons, designated 1989 N2, 1989 N3, and 1989 N4, ranged between 60 and 125 miles in diameter and were found to be in normal equatorial orbits in the direction of the planet's rotation. Prior to Voyager's observations, only two moons of Neptune had been observed. (NASA Release 89-128, Aug 3/89; *NY Times,* Aug 4/89; *W Post,* Aug 4/89; *LA Times,* Aug 4/89;

August 8–13: The Space Shuttle Columbia was successfully launched from the Kennedy Space Center in Florida, on mission STS-28 to deploy a classified Department of Defense payload. This was the first launch of the Columbia orbiter in the three and a half years since the explosion of the Space Shuttle Challenger. Various media sources reported that Columbia had deployed an advanced military reconnaissance satellite and performed tests of Strategic Defense Initiative (SDI) hardware. Columbia landed at Edwards Air Force Base, California, on Sunday, August 13. (SSR 1989 061A; NASA PFOR, M-989-89-28; NASA MOR, E-199-89-28; *UPI,* Aug 8/89; *AP,* Aug 8/89; *NY Times,* Aug 9/89; *W Post,* Aug 9/89; *W Times,* Aug 9/89; *USA Today,* Aug 9/89; *LA Times,* Aug 9/89; *C Trib,* Aug 9/89; *P Inq,* Aug 9/89; *B Sun,* Aug 9/89)

August 8: An ESA Ariane 4 launch vehicle carrying a communications satellite and an orbital observatory was successfully launched from the Kourou space center in French Guiana. The payload consisted of a West German TV-SAT-2 communications satellite and the Hipparcos orbiting star mapper. (SSR 1989 062A-D; *LA Times,* Aug 9/89)

• An attempted launch of a Japanese H-1 launch vehicle was aborted after one of the rocket's main engines failed to ignite. (FBIS-Eas-89-151, Aug 8/89; *WSJ*, Aug 9/89; *C Trib*, Aug 9/89)

August 11: Planetary scientists at NASA's Jet Propulsion Laboratory, Pasadena, California announced the discovery of "ring arcs" orbiting Neptune. NASA scientists noted that, unlike the rings surrounding Saturn, Jupiter, and Uranus, Neptune's rings did not appear to completely surround the planet. (NASA Release 89-132, Aug 11/89; *NY Times*, Aug 12/89; *W Post*, Aug 12/89; *P Inq*, Aug 12/89; *C Trib*, Aug 12/89; *B Sun*, Aug 12/89; *WSJ*, Aug 14/89)

August 18: An Air Force Delta 2 launch vehicle carrying a NAVSTAR satellite was successfully launched from Cape Canaveral Air Force Station in Florida. The 126-foot Delta 2 placed the NAVSTAR navigation satellite in a preliminary orbit, from where it would boost itself into a 12,000-mile circular orbit. The NAVSTAR was the third of an eventual network of 21 satellites that would comprise the Global Positioning System, a highly accurate navigation system for military and commercial users. (SSR 1989 064A-C; *AP*, Aug 18/89; *UPI*, Aug 18/89; *W Post*, Aug 19/89; *LA Times*, Aug 19/89; *W Times*, Aug 21/89)

August 22: While heading toward a close approach of Neptune, the Voyager 2 space probe discovered the first complete ring around the planet. Scientists at NASA's Jet Propulsion Laboratory, Pasadena, California, had previously observed several ring arcs. The discovery of the complete ring around Neptune had been expected based on observations of rings on the other three large outer planets during previous Voyager planetary encounters. (NASA Release 89-132, Aug 11/89; *NY Times*, Aug 23/89; *W Post*, Aug 23/89; *WSJ*, Aug 23/89; *B Sun*, Aug 23/89; *UPI*, Aug 23/89)

August 25: After travelling 4.4 billion miles over twelve years, the Voyager 2 space probe completed its fourth and final planetary encounter by passing Neptune at a distance of 3,042 miles. During the Neptune encounter, Voyager 2 discovered six new moons of Neptune, a set of faint rings around the planet, auroral activity on Neptune and Triton, and a surprisingly dynamic atmosphere containing large storm systems. Additionally, Voyager 2 obtained closeup images of Neptune's two largest moons, Triton and Nereid. Voyager 2 and its sister spacecraft, Voyager 1, were launched in 1977 for planned encounters with Jupiter and Saturn. After its Saturn encounter, Voyager 1 was steered away from the plane of the solar system toward interstellar space, while the Voyager 2 mission was expanded to take advantage of a rare alignment of the large outer planets, which made possible a "grand tour" of the outer solar system. With its power supply and all major mechanical systems still operative, Voyager 2 encountered Uranus in 1986, discovering rings and several new moons around the planet.

Advances in radio receiving technology allowed continued communication with Voyager 2 despite the spacecraft's extremely faint signal. At the conclusion of the Voyager 2 encounter with Neptune, the craft would steer clear of the plane of the solar system and continue traveling into interstellar space. In anticipation of a possible retrieval of the craft by intelligent extraterrestrial beings, both Voyager spacecraft carried plaques bearing hieroglyphic messages that indicated their origins on Earth and contained images of a human male and female in a gesture of greeting. The spacecraft also carried a gold-plated phonographic record of sounds from Earth, including greetings in all of the world's major languages and the sounds of several animal species. (NASA Headline News, Aug 28/89; *Voyager Bulletin/Mission Status Report No. 96,* Oct 4/89; NASA Voyager 2 Neptune Encounter Press Kit, Aug/89; *UPI,* Aug 25/89; *NY Times,* Aug 25/89; *W Post,* Aug 25/89; *WSJ,* Aug 25/89; *USA Today,* Aug 25/89; *W Times,* Aug 25/89; *P Inq,* Aug 25/89; *CSM,* Aug 25/89; *B Sun,* Aug 25/89)

August 28: Astronomers at Cornell University announced the discovery of an apparent galaxy in the making, challenging conventional theory that galaxies formed only during the early history of the universe. The galaxy observed by Cornell astronomer Martha Haynes and Ricardo Giovanelli appeared to contain no fully developed stars and to consist of only protostellar material, mainly hydrogen. The observed galaxy, named UGC-7394, was 65 million light years distant. (*UPI,* Aug 29/89; *WSJ,* Aug 29/89; *W Times,* Aug 29/89; *B Sun,* Aug 29/89; *C Trib,* Aug 29/89)

August 29: The Italian space agency announced that its star mapper satellite, *Hipparcos 1,* failed to boost itself into the proper orbit. *Hipparcos 1* was launched aboard an *Ariane 4* launch vehicle on August 8. The satellite's designers said that they would seek funds from the European Space Agency to build and launch a replacement satellite. (SSR 1989 062 A-D; *NY Times,* Aug 30/89)

September

September 4: The last Titan 34D launch vehicle in the Air Force's inventory was successfully launched from Cape Canaveral Air Force station in Florida, carrying a classified Department of Defense payload. This was the final launch of the Titan 3 series launch vehicles, which were being replaced by the more powerful Titan 4. (SSR 1989 069 A-D; *UPI,* Sep 4/89; *NY Times,* Sep 5/89; *W Post,* Sep 5/89; *USA Today,* Sep 5/89; *C Trib,* Sep 5/89; *LA Times,* Sep 5/89; *W Times,* Sep 5/89)

September 5: After an aborted attempt in August, an H-1 launch vehicle carrying a GMS-4 weather satellite was successfully launched by Japan's National Space Development Agency. (FBIS-Eas-89-173, Sep 8/89; SSR 1989 070 A-C; *WSJ,* Sep 7/89)

September 6: Two Soviet astronauts were successfully launched aboard a Soyuz TM-8 capsule for a rendezvous with the Mir Space Station. The Soyuz TM-8 capsule was launched aboard a Proton launch vehicle emblazoned with several commercial advertisements. Cosmonauts Alexander S. Viktorenko and Alexander A. Serebrov were scheduled to reoccupy the Mir station during a six-month tour and supervise the attachment of two new modules to the facility. Mir had been left unoccupied since April 27 because of funding constraints and delays in manufacturing of the new modules. (FBIS-Sov-89-171, Sep 6/89; SSR 1989 071A; *W Post,* Sep 6/89; *WSJ,* Sep 6/89; *C Trib,* Sep 6/89; *P Inq,* Sep 6/89)

September 7: NASA announced it had awarded Ford Aerospace a $500 million contract to modernize its Mission Control Center at the Johnson Space Center, Houston, Texas, and build a new control center for future Space Station operations. (NASA Release, Sep 7/89; *Reuters,* Sep 7/89; *NY Times,* Sep 7/89; *W Post,* Sep 7/89)

September 8: Two Soviet cosmonauts successfully docked with the Mir Space Station and began activating the facility's systems after a four-month hiatus. (FBIS-Sov-89-173, Sep 8/89; SSR 1989 071A; *AP,* Sep 8/89; *UPI,* Sep 8/89)

September 12: Antinuclear activists in Florida protested the launch aboard the Space Shuttle *Atlantis* next month of the Galileo space probe, which would carry two plutonium-fueled radioisotope thermoelectric generators. The protesters claimed an accident aboard the Atlantis similar to the *Challenger* explosion could contaminate large parts of Florida and lead to hundreds of cancer-related deaths. NASA and the White House maintained that the chances of a plutonium leak in the event of an accident was negligible. Galileo was scheduled to reach Jupiter in late 1995. (*AP,* Sep 12/89; *UPI,* Sep 12/89; *USA Today,* Sep 13/89)

September 24: The Space Shuttle Columbia was accidentally sprayed with water when an automatic sprinkler system went off unexpectedly inside the Kennedy Space Center (KSC) Orbiter Processing Facility No. 2. KSC engineers examined the orbiter for damage and planned an investigation of the incident. (*UPI,* Sep 24/89; *W Times,* Sep 25/89; *USA Today,* Sep 25/89)

September 25: A NASA Atlas-Centaur vehicle carrying a Navy communications satellite was successfully launched from Cape Canaveral Air Force Station, Florida. This was the last orbital launch of a launch vehicle by NASA, since responsibility for all future orbital launches was transferred to the private sector and the branches of the Defense Department. Since inception in 1958, NASA had launched over 400 untended orbital rockets. The Atlas-Centaur carried a Navy Fleet Satellite Communications F-8 (FLTSATCOM F-8) relay station into a preliminary orbit on its way toward geosynchronous orbit. (NASA Release 89-145, Sep 18/89; NASA Release, Sep 25/89; NASA: Atlas/Centaur-68 *FLTSATCOM F-8 Launch Press Kit,* Sep/89; *AP,* Sep 25/89; *UPI,* Sep 25/89; *USA Today,* Sep 25/89; *NY Times,* Sep 26/89; *WSJ,* Sep 26/89; *LA Times,* Sep 26/89; *P Inq,* Sep 26/89; *B Sun,* Sep 26/89; *W Times,* Sep 26/89)

September 27: NASA announced it had granted a $1.3 billion extension of the contract to Lockheed Corporation, Calabasas, California, to continue Space Shuttle processing operations for three more years. (NASA Release, Sep 27/89; *AP,* Sep 26/89; *W Post,* Sep 27/89)

• During a visit by Soviet journalists to the Plesetsk Space Center, the Soviet government revealed that a launch pad explosion of a Vostok rocket had killed 50 people in 1980. The Soviet government also revealed that a Cosmos rocket had also exploded on the launch pad in 1973, killing nine technicians and soldiers. (FBIS-Sov-89-185, Sep 26/89; *NY Times,* Sep 28/89; *W Times,* Sep 28/89; *P Inq,* Sep 28/89)

September 28: A coalition of three antinuclear groups filed suit in the Washington, D.C. Federal District Court to stop the launch of the nuclear-powered Galileo space probe aboard the Space Shuttle Atlantis. The three groups contended that NASA and the Department of Energy had illegally delayed release of health and safety risk assessments of a potential accident involving Galileo's plutonium-fueled radioisotope thermal generators. The groups filing suit were the Florida Coalition for Peace and Justice, the Christic Institute, and the Foundation on Economic Trends. NASA and the Energy

Department denied the charges, and the space agency continued to carry on with preparations for the planned launch of Atlantis-Galileo on October 12. The suit was subsequently dismissed by the court during its preliminary hearing. (*LA Times*, Sep 28/89; *NY Times*, Sep 29/89; *W Post*, Sep 29/89; *W Times*, Sep 29/89; *B Sun*, Sep 29/89; *UPI*, Sep 29/89)

September 29: NASA named 18 astronauts to fly on future Space Shuttle missions, including the first African-American woman and the first member of the Coast Guard to fly aboard the Shuttle. Dr. Mae Jemison, a physician, would become the first African–American woman to fly in space. She was scheduled for the June 17, 1991, Spacelab mission aboard Discovery. (*AP*, Sep 9/89; *UPI*, Sep 29/89; *USA Today*, Oct 4/89; *W Times*, Oct 2/89)

During September: The Soviet news magazine *Izvestia* published an article confirming that the Soviet Union had established a secret program in the 1960s to beat the United States in landing astronauts on the Moon. The Izvestia article claimed that the Soviet Union had secretly conducted a crew-assisted lunar landing program from 1961 to 1969 comparable to the Apollo program and had developed a booster—known as the N1—comparable to the United States Saturn V. The Soviet lunar program was eventually abandoned after three attempts to launch the N1 failed; the United States successfully landed astronauts on the Moon in 1969. (FBIS-Sov-89-210, Nov 1/89; *H Chron*, Sep 17/89)

October

October 2: Planetary scientists at NASA's Jet Propulsion Laboratory, Pasadena, California, announced that Voyager 2 photographs taken during the August encounter with Neptune had revealed the presence of an active volcano or geyser on Neptune's moon, Triton. The photographs revealed a dark plume nearly 5 miles high and 90 miles in length emanating from the surface of Triton. Triton became only the third object in the solar system known to have active eruptions from its surface. (NASA Release 89-156, Oct 2/89; *Voyager Bulletin/Mission Status Report* No. 95, Oct 2/89; *AP*, Oct 3/89; *UPI*, Oct 3/89; *NY Times,* Oct 3/89; *USA Today,* Oct 3/89; *P Inq,* Oct 3/89; *C Trib,* Oct 3/89)

October 5: The first attempt to launch a fully private funded rocket ended in failure when the 58-foot-tall "Koopman Express" sounding rocket caught fire while still on the launch pad. The rocket was to have flown on a 15-minute suborbital flight from Vandenberg Air Force Base, California, during which it would have released a cloud of inert gas to test sensors on a military Delta Star satellite. The rocket was also scheduled to deploy an engineering model of a Parashield during reentry into the atmosphere. (*NY Times,* Oct 6/89; *W Post,* Oct 6/89; *USA Today,* Oct 6/89; *W Times,* Oct 6/89; *B Sun,* Oct 6/89)

October 18–23: The Space Shuttle Atlantis was successfully launched from Kennedy Space Center (KSC), Florida, on mission STS-34 to deploy the $1.4 billion Galileo Jupiter space probe. The Atlantis launch had been delayed by a week because of problems with a computer, a main engine master control device, and, finally, rain. Shortly after reaching orbit, Atlantis released the Galileo probe and the Inertial Upper Stage (IUS) from the cargo bay. The IUS subsequently fired Galileo on a trajectory toward Venus, in preparation for a gravity assist maneuver that would swing the spacecraft back toward Earth twice and onward to Jupiter. Galileo was scheduled to reach Jupiter in December 1995, whereupon it would begin orbiting the planet for 22 months of observations of the Jovian system. Approximately five months before reaching Jupiter, Galileo would release a 34-inch unpowered descent module that would enter the Jovian upper atmosphere and descend several hundred miles before being destroyed by extreme atmospheric pressure. The Atlantis-Galileo mission had sparked protests and a law suit by anti-nuclear groups concerned over the danger of possible radioactive contamination from the probe's radioisotope thermal generators in the event of a catastrophic accident. In anticipation of possible efforts to disrupt the launch, NASA increased security measures at KSC. In addition to launching Galileo, the STS-34 crew would conduct observations of the Earth's ozone layer and perform several onboard medical, biological and materials processing experiments. The crew included Navy Captain Donald E. Williams (Commander), Navy Commander Michael J. McCulley (Pilot), and Mission Specialists Shannon Lucid, Franklin Chang-Díaz, and Ellen S. Baker. The mission ended successfully on October 23 with Atlantis landing at Edwards Air Force Base, California. (NASA MOR E-829-34-89-01; NASA PFOR M-989-89-34; NASA *STS-34 Press Kit*, Oct/89; NASA Facts: STS-34; Galileo/KSC Release 79-89, Sep/89; SSR 1989 084 A-D; *UPI*, Oct 19/89; *AP*, Oct 19/89; *NY Times*, Oct 19/89; *W Post*, Oct 19/89; *USA Today*, Oct 19/89; *WSJ*, Oct 19/89; *P Inq*, Oct 19/89; *W Times*, Oct 19/89; *B Sun*, Oct 19/89)

October 19: Soviet and western planetary scientists announced results from the Phobos 2 mission to Mars and the moon Phobos. Despite the premature loss of spacecraft transmissions because of a computer error, scientists revealed that valuable data had been obtained from the two-month period preceding the loss of communications. The new findings showed that Mars has no magnetic field of its own and that incoming solar particles impact directly with the planet's ionosphere. Researchers hypothesized that the direct interaction between the solar wind and Mars' upper atmosphere may have been responsible for the depletion of the planet's original atmosphere, as ionospheric particles were gradually electrified by the incoming solar particles. The observations suggested that Mars may have had an atmosphere comparable to Earth's during the first billion years after planet formation, and that conditions for the appearance of life may have prevailed during that time. (*NY Times*, Oct 19/89)

October 21: An Air Force Delta 2 launch vehicle was successfully launched from Cape Canaveral Air Force Station, Florida, carrying the fourth NAVS-TAR navigation satellite. The NAVSTAR would become part of an eventual 21-satellite Global Positioning System (GPS) navigation network. (SSR 1989 085 A-C; *NY Times*, Oct 22/89; *W Times*, Oct 23/89)

November

November 7: The European Space Agency announced the Hipparchos 1 star mapper satellite had been reprogrammed to function in its current orbit. Hipparchos became stranded in a lower than expected orbit after its orbital booster rocket failed to ignite in August. (*P Inq*, Nov 8/89)

November 9: While on the way to a gravity assist flyby of Venus, the Galileo Jupiter space probe made the first course corrections by briefly firing its thrusters. This was the first of about 30 course corrections to be undertaken by Galileo during its five-year trip to Jupiter. (NASA *STS-34 Press Kit*, Oct/89; *NASA Facts: STS-34 Galileo*-KSC Release 79-89, Sep 89; *W Times*, Nov 10/89; *P Inq*, Nov 10/89; *UPI*, Nov 14/89)

November 16: Astronomers at the Harvard-Smithsonian Center for Astrophysics announced the discovery of the largest structure seen so far in the universe. The scientists claimed to have discovered a sheet of galaxies 500 million light-years long, 200 million light years wide, and about 15 million light years thick, which they are calling the "Great Wall". The Great Wall, which is invisible to the naked eye, was estimated to be 200 million to 300 million light-years away from Earth and was expected to reveal new insights on the distribution of matter in the universe. (*UPI*, Nov 17/89; *P Inq*, Nov 17/89; *B Sun*, Nov 17/89)

• The Colorado-based Energetics Satellite Corporation announced that it had completed an agreement with the Soviet Union's civilian space agency, Glavkosmos, to launch two U.S.-built commercial satellites on Soviet boosters. The contract provided for the launching of two geosynchronous satellites at a cost of $6.5 million each, with an option to add six more satellite launches in the future. The total potential value of the agreement was $54 million for the eight-satellite launch option. (*W Post*, Nov 16/89; *NY Times*, Nov 17/89)

November 18: A NASA Delta launch vehicle was successfully launched on this date from Vandenberg Air Force Base, California, carrying the $160 million Cosmic Background Explorer (COBE) satellite into a 570-mile polar orbit. Three instruments aboard COBE would record the background radiation left over from the "Big Bang" in unprecedented detail, allowing refinement of theories on the formation of galaxies and galactic clusters in the early universe. The Delta was the last launch vehicle in NASA's inventory. Hereafter all future NASA untended launches would occur on commercial or military boosters. (NASA Release N89-74; NASA *Cosmic Background Explorer (COBE) Press*

Kit Release 89-172; SSR 1989 A-D; AP, Nov 18/89; *UPI*, Nov 18/89; *LA Times*, Nov 18/89; *P Inq*, Nov 19/89; *C Trib*, Nov 19/89; *CSM*, Nov 20/89)

November 19: A team of astronomers from the California Institute of Technology (Caltech) and Princeton University announced the discovery of the oldest and most distant object yet observed in the universe. The object was reported to be a quasar at a distance of 10 to 15 billion light-years from Earth, which may have formed when the universe was only about a billion years old. The discovery challenged conventional theories on the timetable for the formation of galaxies and galactic-sized bodies in the early universe. Previous observations had placed the age of most quasars at about 3 billion years after the Big Bang. The new observations were expected to produce major revisions in theories of how matter coalesced during the early universe. (*NY Times*, Nov 20/89; *W Post*, Nov 20/89; *W Times*, Nov 20/89; *P Inq*, Nov 20/89; *B Sun*, Nov 20/89)

November 22–27: The Space Shuttle Discovery was successfully launched at night from Kennedy Space Center in Florida, on mission STS-33. Discovery carried a classified Department of Defense payload believed to be a $300 million surveillance satellite. The satellite was deployed from the Shuttle's cargo bay on November 23, and was scheduled to boost itself into a geostationary orbit. Several experiments of Strategic Defense Initiative (SDI) technology were also expected to be performed. Crew members for STS-33 were Air Force Colonel Frederick Gregory (Commander), Air Force Colonel John Blaha (Pilot), and Mission Specialists Navy Captain Manley L. Carter Jr., F. Story Musgrave, and Kathryn C. Thornton. Discovery landed at Edwards Air Force Base, California on November 27. (NASA Release 89-75, Nov 15/89; SSR 1989 090 A; *AP*, Nov 23/89; *UPI*, Nov 23/89; *NY Times*, Nov 23/89; *W Post*, Nov 23/89; *WSJ*, Nov 23/89; *W Times*, Nov 23/89; *P Inq*, Nov 23/89)

November 26: The Soviet Union successfully launched an untended Proton launch vehicle carrying an attachment module bound for the Mir Space Station from the Baikonur Cosmodrome, Kazakhstan. The pressurized module, known as the Kvant-2, contained additional crew living space as well as experiments and a "space motorcycle" intended for use in extravehicular spacewalks. (FBIS-Sov-89-226, Nov 27/89; SSR 1989 093 A; *NY Times*, Nov 27/89; *C Trib*, Nov 27/89)

December

December 2: The Solar Maximum (Solar Max) satellite tumbled out of orbit and reentered the Earth's atmosphere, spreading small amounts of debris over the Indian Ocean. The satellite had been launched in February 1980 to conduct gamma ray and ultraviolet observations of the Sun, the Earth's atmosphere, and deep space objects. In April 1984, Solar Max was retrieved by the Space Shuttle Challenger (Mission 41-C) for in-orbit repairs and redeployment. Solar Max's orbit had been deteriorating faster than expected over the past several months because of unusually high solar activity, which caused the Earth's atmosphere to expand and increase the amount of aerodynamic drag experienced by the satellite. (SSR 1980 014 A; *AP*, Dec 2/89; *UPI*, Dec 2/89; *NY Times*, Dec 3/89; *W Post*, Dec 3/89; *P Inq*, Dec 3/89; *B Sun*, Dec 3/8)

December 5: Iraq launched a three-stage rocket capable of putting a satellite into space on a suborbital trajectory from the Al Anbar base in northern Iraq. The launch of the Iraqi rocket was confirmed by the U.S. and Israeli governments. (FBIS-Nes-89-235, Dec 8/89; *UPI*, Dec 9/89; *NY Times*, Dec 9/89)

• Planetary scientists working with data from a Voyager 2 encounter with Neptune announced that the spacecraft had recorded wind speeds on the planet of up to 1,500 miles per hour. These wind speeds exceeded those observed on Saturn and were believed to be the highest in the solar system. The scientists also reported the discovery of at least two additional volcanic plumes on Neptune's moon, Triton. (*AP*, Dec 6/89; *C Trib*, Dec 6/89; *NY Times*, Dec 7/89)

December 11: An Air Force Delta 2 launch vehicle was successfully launched from Cape Canaveral Air Force Station in Florida, carrying a NAVSTAR navigation satellite. The NAVSTAR would be the fifth in an eventual network of 21 military and commercial satellites that would comprise the Global Positioning System. (SSR 1989 097 A-C; *AP*, Dec 11/89; *NY Times*, Dec 12/89; *USA Today*, Dec 12/89; *W Times*, Dec 12/89; *C Trib*, Dec 12/89)

• NASA Administrator Richard H. Truly named Dale L. Compton to be Director of the NASA Ames Research Center, Moffett Field, California, effective December 20. Compton, who succeeded William F. Ballhaus, had been Acting Director since Ballhaus resigned on July 15. (NASA Release 89-182, Dec 11/89)

December 13: Astronomers from the National Optical Astronomy Observatories, Tucson, Arizona, announced the discovery of the furthest known star in the Milky Way Galaxy. The star, which is believed to be 160,000 light years from Earth, was accidentally discovered while the astronomers were conducting a survey of the Virgo cluster of galaxies. The discovery was expected to help determine the mass of the Milky Way and other similar galaxies. (*NY Times*, Dec 14/89; *B Sun*, Dec 14/89)

December 17: Engineers from the Massachusetts Institute of Technology returned from a visit to the Soviet Union with photographs of a crew-assisted lunar landing vehicle that was to be used to land Soviet cosmonauts on the Moon in 1968. The westerners were given the first concrete evidence of a Soviet lunar landing program when they were shown components of a lunar lander at the Moscow Aviation Institute and were allowed to take photographs. Soviet officials confirmed that a race to land astronauts on the Moon had indeed occurred during the 1960s but that the Soviets had fallen behind the United States as a result of several launch failures of their N1 rocket. At the time of the successful lunar mission of Apollo 11, the Soviet Union had denied having a crew-tended lunar landing program. (*UPI*, Dec 18/89; *NY Times*, Dec 18/89; *USA Today*, Dec 18/89; *W Post*, Dec 18/89; *W Times*, Dec 18/89; *P Inq*, Dec 18/89; *C Trib*, Dec 18/89; *B Sun*; Dec 18/89)

December 20: An untended Soviet Progress M-2 module bound for the Mir Space Station was successfully launched from the Baikonur Cosmodrome in Central Asia. In addition to ferrying supplies to the Mir crew, the Progress also carried a U.S. experiment designed to test crystal growth in weightlessness. (FBIS-Sov-89-243, Dec 20/89; SSR 1989 099 A; *W Post*, Dec 21/89; *WSJ*, Dec 21/89; *P Inq*, Dec 21/89; *W Times*, Dec 21/89; *CSM*, Dec 22/89)

January

January 1: The Johnson Space Center in Houston, Texas, awarded Rockwell International Corporation a five-year extension on the contract for systems integration. The $1.36 billion contract extensions were for flight and ground systems engineering maintenance and safety analysis, including configuration of the hardware required for each orbiter flight and Shuttle mission support from countdown to landing. (NASA Release C90-A; C90-B; *WSJ,* Jan 12/90)

January 3: Technical assistant Edwin Sherry, from the Jet Propulsion Laboratory in Pasadena, California, explained a minor problem that the Spacecraft Magellan had developed three days earlier and a little more than halfway into its journey to map Venus. A computer chip failure, caused either by electrical corrosion or an electrically charged particle spewed out by the sun, necessitated frequent commands from ground crews to keep Magellan moving in the right direction. The craft's guidance system normally focused on two stars to ascertain and adjust its heading. Sherry explained that some difficulties with the vast array of electronics during flight were expected and that the mission was in no danger. Scientists hoped to send a new program aloft that would allow the ship's computer to bypass the damaged chip. (*AP,* Jan 4/90)

• NASA released a request on this date to various companies for a proposal concerning design, management, building and testing of the Advanced Tracking and Data Relay Satellite System. The first ATDRSS was scheduled for delivery in 1987 and could replace one of the older TDRS systems in orbit. By 2001, "A four-satellite constellation of ATDRSS is expected to be in orbit," said Thomas Underwood, Assistant Chief for the TDRS system at Goddard Space Flight Center, Greenbelt, Maryland. The ATDRSS promised to offer a much higher communication rate needed for Space Station Freedom missions. (NASA Release 90-2; C90-x)

January 5: NASA Administrator Richard Truly named Thomas E. Utsman as Deputy Associate Administrator for Space Flight. He was to assist William Lenoir, Associate Administrator, in daily management of space flight programs. Utsman began at NASA in 1963 as a facilities design engineer in the Apollo program and later was named to head the Design Engineering Directorate's Project Engineering Office. He also served as Deputy Director of Project Management, Associate Director of Design Engineering, Operations Management Director of Technical Support, Director of Shuttle Operations, and Director of Shuttle Management and Operations.

Truly also named James A. Thomas as Deputy Director of the Kennedy Space Center at Cape Canaveral, Florida. Thomas joined NASA in 1962 and served as lead engineer for pre-launch testing and communications for the Apollo program. He worked on the initial Shuttle program and served as lead flight engineer, as chief Shuttle flight project engineer, in the Shuttle Engineering Directorate, and as Director of Shuttle Launch and Landing Operations. (NASA Release 90-3)

January 8: NASA and Japan's Institute of Space and Astronautical Science agreed to launch Japan's Geotail spacecraft on a Delta II rocket from Kennedy Space Center, in Florida, in 1992. The craft was to study the stored energy in the geomagnetic tail of the Earth, a magnetic field drawn by the solar winds on the Earth's night side and responsible for the Aurora Borealis. (NASA Release 90-4)

January 9: NASA launched the first of 10 expected Shuttle flights in 1990 (flight STS-32) from Cape Canaveral, Florida, with a 10-day mission to orbit a Navy communications satellite and to retrieve another. The liftoff, originally scheduled for December 18, had been delayed three times because of problems with the renovated launch pad and once because of low clouds. The Navy satellite was deployed January 10. On January 12, the spacecraft closed in on and retrieved the Long Duration Exposure Facility (LDEF), an 11-ton satellite, into the cargo bay, using a 50-foot robotic arm.

The LDEF was sent up with Space Shuttle Challenger in April 1984 to provide vital information on the design of spacecraft and NASA's Space Station, as well as other experiments. The orbiting laboratory was scheduled to be recovered 10 months after its deployment, but the nearly five-year postponement caused by scheduling problems and the Challenger accident resulted in a decaying orbit that was expected to reach Earth's atmosphere by March of 1990. The 57 experiments aboard the 30-foot-long cylindrical satellite included an array of potential space building materials being tested for reactions to radiation, extreme temperature changes, and collisions with space matter. Also tested for prolonged space exposure were optical fibers, pure crystals for use in electronics, and a student experiment involving tomato seeds. Interstellar gases and cosmic radiation were also trapped in an attempt to find clues into the formation of the Milky Way and the evolution of heavier elements. For the remainder of the mission, astronauts carried out other scientific and engineering work.

Columbia landed on a concrete runway early in the morning on January 20 at Edwards Air Force Base in California after a landing postponed for a day because of fog. The flight set a precedent both because of duration—11 days—and because the aircraft was the heaviest Shuttle ever to land, a result of the captured cargo. The mission was without incident, save for a few minor problems involving a leaky dehumidifier, a smoke detector that went off unexplainably, a malfunction in one of the three Inertial Measurement Units (critical for reentry, but the Shuttle can land with only one functioning unit), and a small engine used by the automatic pilot for maintaining flight path that fired for no reason. The retrieved satellite remained in the Shuttle cargo bay and was scheduled to be flown to Kennedy Space Center in Florida atop a modified NASA jumbo jet later in the week.

During re-entry, two NASA experiments measured the aerodynamic and thermodynamic characteristics of the orbiter. One of the experiments, a Shuttle Infrared Leeside Temperature Sensing experiment, involved an infrared camera used to collect imagery from the left wing, an area that reached 1,100 degrees Fahrenheit, a higher than usual temperature, on the STS-28 flight. The second experiment involved the Shuttle Entry Air Data System, housed in Columbia's nose cap to gather "air data" to determine the orbiter's aerodynamic flight characteristics.(*NY Times,* Jan 10/90; Jan 11/90; Jan 13/90; Jan 21/90; *W Post,* Jan 10/90; Jan 11/90; Jan 13/90; Jan 21/90; *C Trib,* Jan 13/90; *CSM,* Jan 16/90; NASA Release 90-8)

January 10: Vice President Dan Quayle, leader of the White House National Space Council, told scientists at the 175th meeting of the American Astronomical Society that "the rest of the world is catching up and may pass us by" because our space programs consume too much time and money. His eight-page speech was not lauded by the audience until he suggested greater emphasis on untended space flight; many scientists, dependant on Federal funding, view crew-assisted flights as a drain on·their own projects. Quayle reiterated a commitment to balance tended and untended space exploration and called upon allied nations, most specifically Japan, and the business sector to share costs.

The Vice President also asked the NASA to look "across the traditional divisions among civil, commercial, and national security activities," which some NASA sources interpreted as a suggestion that outside agencies might assume a bigger role in the space program. In accord with Quayle's comment concerning lengthy and costly space projects, John Pike, Associate Director for Space Policy at the Federation of American Scientists, brought up NASA plans for crew-assisted flights to the Moon and to Mars. The missions, he noted, would cost $15 to $20 billion annually over 25 years while guaranteeing job security for NASA. (*P Inq,* Jan 11/90; *W Post,* Jan 11/90; *NY Times,* Jan 11/90, *W Times,* Jan 11/90)

January 13: NASA scientists reported to the American Astronomical Society preliminary information obtained from the Cosmic Background Explorer spacecraft, launched on November 18, 1989, aboard a Delta rocket. The data gleaned by the Far Infrared Absolute Spectrophotometer, the Differential Microwave Radiometer, and the Diffuse Infrared Background Experiment supported the Big Bang theory, but at the same time minimized the belief that other releases of energy had subsequently occurred. Scientists reported that instruments aboard the craft performed with unprecedented precision. This precision was predicted to increase over the spacecraft's estimated two more years of data gathering. (NASA Release 90-5; *P Inq*, Jan 15/90)

January 15: The White House began looking at other ideas for astronaut expeditions to the Moon and Mars besides those presented by NASA. The unprecedented move stemmed from NASA's cost and time estimate, $400 billion over 30 years for the project. A panel representing 54 aerospace companies had met in the previous week, and the National Research Council planned to review its own efforts to seek innovative ideas later in the week. NASA joined the search for sources of outside ideas as well. Arnold Aldrich, NASA Associate Director for Aeronautics, called the move "highly appropriate" and added, "A lot of our ideas are good, but we certainly haven't covered the waterfront." (*NY Times*, Jan 15/90)

January 17: NASA selected 23 new astronaut candidates from a list of 1,945 qualified applicants for one year of training and evaluation that was to begin in July 1990. Of 11 civilians and 12 military officers, 7 were selected as pilots and the other 16 as mission specialists. (NASA Release 90-7)

January 19: NASA announced the selection of two payload specialists for the International Microgravity Laboratory mission aboard the Space Shuttle Columbia, scheduled for launch in December 1990. Dr. Roberta Bondar of the Canadian Space Agency and Dr. Ulf Merbold of the European Space Agency were to conduct the first of several microgravity investigations using the Spacelab Module. (NASA Release 90-9)

January 21: The European Space Agency successfully sent up an Ariane 4 rocket from its launch site in French Guiana and its payload, a SPOT 2 observation satellite and six smaller satellites, that included four tiny U.S. spacecraft for education and science. It was the first 1990 launch for ESA, the thirty-first since 1979. All seven satellites were successfully deployed. (*UPI*, Jan 22/90)

January 24: Four major U.S. defense contractors—General Dynamics Corporation, McDonnell Douglas Corporation, United Technologies Corporation, and Rockwell International Corporation—announced they would jointly develop the X-30 National Aerospace Plane. Until September 1989, industry had roughly matched Government investment in the project, $806.3 million since inception during the Reagan administration. "Fiscal realities," said the *Wall Street Journal*, "spur the call for teamwork". (*W Post*, Jan 24/90; *WSJ*, Jan 24/90)

• Japan launched a lunar probe from Uchinoura, on their southern coast, making it the third nation, after the United States and the Soviet Union, able to send up a moon probe. The flight schedule was to go into elliptical orbit around the Earth until it came within 10,000 miles of the Moon, launch an instrument package to orbit the moon, and send back data concerning temperatures and electrical fields. (*WSJ*, Jan 25/90; *W Post*, Jan 25/90; *NY Times*, Jan 25/90)

• Johnson Space Center, Houston, Texas, issued a call for proposals for design of a Satellite Servicer System flight demonstrator to show the ability to maintain satellites in polar and high inclination orbits. The on-orbit flight demonstration would exercise rendezvous and docking, as well as orbital replacement unit exchange and fluid transfer. (NASA Release 90-12)

January 25: The projected cost for NASA's Earth Observing System (EOS) had soared from an estimated $12 billion to $30 billion dollars, Administration sources announced. It was the largest and most expensive project yet undertaken by NASA. President George Bush had asked the National Academy of Sciences to review the matter and recommend cost reductions.

Lauded by environmental scientists and termed by former astronaut Sally Ride as a "mission to planet Earth," the plan called for putting platforms, launched by Titan 4 rockets from Vandenberg Air Force Base in California, into polar orbit. Each platform would carry 10 sophisticated scientific instruments that could send 10 trillion bytes of data daily. The first EOS platform, said NASA, would not be in orbit until 1998. (*LA Times*, Jan 26/90)

January 26: NASA granted a small Alexandria firm, Global Outpost, rights to find ways of using empty Space Shuttle fuel tanks, used to carry 500,000 gallons of liquid hydrogen and oxygen for Space Shuttle launch and initial insertion into orbit, as a platform for military, civilian, and commercial experiments. The agreement followed a June 1988 NASA announcement requesting commercial and academic uses for expended tanks. (*W Times*, Jan 29/90; NASA Release 90-14)

January 28: President George Bush prepared to send a 1991 budget spending plan to Congress that proposed a 24 percent increase for NASA, bringing its fiscal year 1991 budget to $15.12 billion. The increased funds were earmarked for the Space Station, a satellite program to monitor Earth's environment (EOS), and another for sending astronauts to the Moon and Mars. (*WSJ*, Jan 29/90)

January 29: NASA announced a recently revised 1990 schedule for Space Shuttle launches that called for 9 flights instead of the previously announced 10. The reason behind the shuffling hinged on a three-week delay to launch the Hubble Telescope, originally planned for deployment March 26, because of an unsatisfactory leak check of the Space Shuttle's solid rocket booster. This delay threatened the high-priority Ulysses mission scheduled for October 5. The mission had to be sent up within an 18-day time period in order to slingshot Jupiter and head back toward the sun, where it would achieve polar orbit around the star. The new agenda for Shuttle launches and their payloads is as follows:

- Jan 9: Columbia; Navy communications satellite and retrieval of another.

- Feb 22: Atlantis; classified military satellite.

- Apr 18: Discovery; the Hubble Telescope.

- May 9: Columbia; "astro" astronomy telescopes.

- Jul 9: Atlantis; classified military cargo.

- Aug 29: Columbia; a Spacelab life sciences module.

- Oct 5: Discovery; the Ulysses probe.

- Nov 1: Atlantis; Gamma Ray Observatory.

- Dec 12: Columbia; a materials science laboratory.

A military mission carrying a variety of infrared sensors for the Strategic Defense Initiative was reassigned for 1992. NASA planned for 64 Space Shuttle flights and 30 expendable launch vehicle missions through September 1995. (*UPI*, Jan 29/90; *W Times*, Jan 30/90; *NY Times*, Feb 4/90; NASA Release 90-15)

January 30: The Sandia National Laboratories in New Mexico issued a report concerning an electric gun, the coil gun, that could thrust payloads into orbit. Hoping one day that the device would launch small satellites, Bill Cowan, director of the project, predicted that objects could be hurled into space at 2.8 miles per second at one percent of the cost of conventional rockets. The gun accelerates projectiles by pulling them through successive, interacting magnetic fields created by electric currents passing through coils of wire. According to Cowan, obstacles such as damage to the payloads and launchers themselves and the heat generated by projectiles rapidly propelled through the atmosphere would eventually be overcome. (*NY Times*, Jan 30/90)

- Astronauts who retrieved the Long Duration Exposure Facility in early January said, in a press briefing, that the satellite was in worse shape than engineers had expected after its nearly 6-year orbit. The conclusion was that materials used for the planned Space Station would require greater scrutiny. (*W Times*, Jan 31/90)

January 31: Columbia's flight in early January, NASA announced, was made with a flawed backup o-ring inspectors had overlooked. A disaster similar to the Challenger accident four years ago could have occurred if the primary seal had failed. Inspectors also missed inferior welds on a critical pump that supplied one of the Atlantis main engines with fuel. The problem was corrected while on the launch pad in preparation for the upcoming February 22 launch. (*Orl Sen,* Feb 1/90; *Fla Today,* Feb 1/90)

During January: NASA completed a facility to house the second ground station for its Tracking and Data Relay Satellite (TDRS) system. Located at White Sands, New Mexico, near the existing terminal, the station was predicted to be fully operational by 1993. The equipment that the new terminal would house was to be technically superior to the communications network in the original building, but the two were to work together in a combination meant to reduce the likelihood for any loss of contact with the two TDRS satellites in orbit. (*CSM,* Jan 30/90)

• NASA signed an agreement with the Maryland Department of Economic and Employment Development for technology sharing designed to increase commercial exploitation of space technology. The agreement, like the ones reached earlier with Virginia and West Virginia, would mainly involve supplying NASA research reports, sponsoring seminars and lectures, and making NASA laboratories available to the private sector. (*W Post,* Jan 5/90)

February

February 1: NASA announced the appointments of Peter G. Smith to the position of Director, International Relations Division, and David L. Stottlemyer to the position of Director, Industry Relations Division. Smith, who would direct NASA's internal relations, worked with the Department of State before coming to NASA in 1979, where he was Chief of the International Program for Policy and Deputy Director of International Relations. Stottlemyer previously held a number of positions with the Foreign Service and the United Nations. (NASA Release 90-20)

February 4: On this date, China put its fifth relay satellite into space on a Long March 3 rocket sent up from the Xichang center in western China and announced that it was ready to launch foreign satellites. China had scheduled launches for Hong Kong and Australian companies who wanted to send up three American-built satellites beginning in April of 1990. (Beijing Xinhua domestic service in Chinese, Feb 20/90; *W Times*, Feb 5/90)

February 5: NASA selected 28 research proposals for negotiation of Phase II contract awards in their Small Business Innovation Research program. The awards were worth $13 million and part of an ongoing effort to stimulate technological innovation, increase the use of the small business sector, and increase the commercialization of federally funded research. (NASA Release 90-21)

February 6: NASA Administrator Richard Truly addressed the House Science, Space, and Technology Committee and defended the projected 24 percent budget increase that President George Bush had proposed for the agency. Truly warned that any attempt to slash the budget would render the Space Station defunct, a project whose expected cost had risen from the original 1984 estimate of $8 billion to $20 billion, despite reduction in design. Also discussed was the status of the planned Shuttle-C, an crew-assisted heavy-lift cargo rocket that representatives claimed could be built by private industry. It was doubtful that the rocket could be ready in time to aid construction of the Space Station. William Lenoir, NASA Associate Administrator, said the new rocket was not necessary for building the Space Station, but would have other uses.

In a related matter, Truly released to the White House and Congress his plan to increase NASA's budget by 60 percent over the next three years: $15 billion in 1991, $17.6 billion in 1992, $19.3 billion in 1993. The increases, he said, were needed for the Space Station Freedom, the Earth Observing

System, and crew-assisted Moon and Mars explorations. (Hearings before the Subcommittee on Space Science and Applications, Feb 6, 8, 21, 27/90; *Sp News,* Feb 15–11/90; *H Chron,* Feb 7/90; *UPI,* Feb 6/90)

February 10: The Galileo spacecraft launched from the Space Shuttle Atlantis on October 18, 1989, looped 10,000 miles above the planet Venus in what NASA engineers termed a "gravity assisted maneuver" that increased its speed by almost 5,000 mph for its trek to Jupiter. After taking 16 pictures of Venusian cloud and wind patterns while searching for lightning, the craft was scheduled to take more pictures of the planet during its return journey. The second set of photographs, which were to be coupled with sensors, were to detect dust, infrared and visible light, and electrically charged particles. Scientists determined that the spacecraft would be within range of the Earth and able to send back data, in October, using its small antenna. It was expected to reach Jupiter in December 1995. (*P Inq,* Feb 12/90; *B Sun,* Feb 11/90; *NY Times,* Feb 11/90)

February 13: Scientists, after taking their first long look at the Long Duration Exposure Facility recovered by Space Shuttle Columbia in January, said the varied surfaces of the satellite showed signs of pits, scars, and discoloration in some areas while being strangely unscathed in others. The surfaces revealed no "showstoppers" however. The satellite was an experiment meant to test materials for use in building spacecraft and the Space Station. (*W Post,* Feb 13/86; NASA Release 90-23)

• The 12-year-old Voyager 1, into deep space and some 3.7 billion miles from Earth on this date, began taking 64 pictures of our solar system for a kind of "family portrait." NASA considered having the pictures, which were too large for publication, displayed in a 100-foot exhibit produced by Los Angeles artist David Hockney. (*USA Today,* Feb 14/86, *AP,* Feb 14/86)

• Marshall Space Flight Center in Huntsville, Alabama, announced the selection of Teledyne Brown Engineering for negotiations leading to a $172 million contract for payload mission integration. Under the contract, the company would provide management, personnel, equipment, services, supplies, facilities, and materials for Shuttle/Spacelab missions. Spacelab, carried in the Shuttle's cargo bay, converted the Shuttle into a versatile on-orbit research center. (NASA Release C90-f)

February 14: A Delta rocket carried two satellites into space from Cape Canaveral, Florida, to test laser technology planned for the Strategic Defense Initiative program. One satellite, the Relay Mirror Experiment, was to reflect a beam fired from a mountain top in Hawaii back to another nearby point on Earth using a mirror. The other satellite, the Low Power Atmospheric Compensation, was to correct laser strike distortions caused by dust, moisture, and gases. The media generally opined that success of the experiment might induce Congress not to slash the $4.5 billion that President George Bush had requested for further research into the project in budget year 1991. (*W Times*, Feb 15/90; *NY Times*, Feb 15/90)

February 18: For the first time, Soviet defense committee lawmakers visited the Kennedy Space Center in Florida as NASA prepared Shuttle Atlantis for a classified satellite launch that, said sources, was to take reconnaissance photographs and intercept military and diplomatic communications. The 12-day visit reflected a new openness in the Soviet Union, and the committee chairman, referring to the mission said, "We have no special secrets...I think you are wasting your money." (*W Post*, Feb 19/90; *NY Times*, Feb 20/90)

February 19: Soviet cosmonauts ended a 22-week mission when their Soyus TM-8 capsule touched down on the Kazakhstan Steppe, southeast of Moscow. The fifth mission involving a crew living aboard the 20-ton Mir Space Station included dozens of scientific experiments and testing of a "space motorcycle," a one-man vehicle designed for servicing satellites. (FBIS Moscow, Tass (Eng trans) Feb 19/90; *NY Times*, Feb 20/90; *P Inq*, Feb 20/90)

February 20: NASA signed an agreement to support the University Corporation for Atmospheric Research (UCAR) study of the feasibility of using the Shuttle's external tanks as research, storage, or manufacturing facilities. Space Shuttle fuel tanks were used to carry 500,000 gallons of liquid hydrogen and oxygen for launch and initial insertion into orbit. The agreement followed a June 1988 NASA announcement requesting commercial and academic uses for expended tanks. (NASA Release 90-26)

• A German scientific satellite, ROSAT, scheduled for NASA launch on a Delta II rocket, in early summer, arrived at Cape Canaveral, Florida, on this date from the Federal Republic of Germany aboard a 747 cargo plane. The spacecraft was to perform "all sky" surveys with two x-ray imaging telescopes, after which it would focus on detailed x-ray sources. (NASA Release 90-27)

February 21: Launching of the Shuttle Atlantis and deployment of a classified payload mission on February 22 were canceled because the mission Commander had a sore throat and headache, coupled with the prediction of wind and rain over Cape Canaveral, Florida. This was the first time a NASA mission had been scrubbed because of crew illness. (*NY Times*, Feb 22/90; Feb 23/90; *W Post*, Feb 22/90)

• Space Shuttle mission specialist Colonel Richard M. Mullane announced his retirement from NASA, effective July 1, 1990. He flew two Shuttle flights in 1984 and 1988 and was scheduled to fly aboard the Atlantis later in February. (NASA Release 90-28)

February 22: An Ariane rocket carrying two American-built Japanese satellites exploded over the Atlantic Ocean shortly after liftoff from French Guiana; a fire in the rocket's propulsion system was suspected to be the cause of the explosion. Ariane had a record of 17 successful flights previous to this accident; it held more than 50 percent of the international market for commercial launches. (The European Space Agency later admitted that a cloth inadvertently left in the rocket had caused the failure by blocking water supplies to one of its first stage engines.) (FBIS Paris AFP in English, Apr 13/90; *W Times,* Feb 23/90; *NY Times,* Feb 23/90)

February 23: Pioneer 11, launched April 5, 1973, was 2.8 billion miles from Earth, passing the planet Neptune (Neptune and Pluto exchange outermost planet status because of Pluto's eccentric orbit) to become the fourth spacecraft that had left the solar system. Its main mission was to glean data from Saturn and Jupiter, but scientists hoped that Pioneer 11 would send back information concerning solar winds until 1995. *(C Trib,* Feb 25/90; *NY Times,* Feb 25/90; NASA Release 90-29)

February 26: Space Shuttle pilot Donald E. Williams announced his retirement from NASA and the Navy, beginning March 1. He piloted two Shuttle missions, in April 1985 and October 1989, was Chief of the Astronaut Office Mission Support Branch, and Deputy Manager of Operations Integration. (NASA Release 90-31)

February 27: NASA issued a request for proposals to the commercial launch industry for launch of up to 10 small Explorer-class spacecraft between 1993 and 1998. The request called for furnishing the launch vehicle, facilities, personnel, and services necessary for the required orbit. (NASA Anno)

February 28: After five delays because of the Shuttle Commander's head cold, a faulty ground computer, and inclement weather, Atlantis (STS-36) lifted off under dark skies from Cape Canaveral, Florida, to deliver a 37,300-pound classified military satellite into orbit. The postponements tacked on an additional $2.7 million to the mission, but NASA Administrator Richard Truly reiterated NASA's policy that meeting launch schedules was secondary to safety observation. The Space Shuttle deployed its $500 million military payload the following day at an orbit inclined 62 degrees north of the equator, allowing it to fly over every major populated and industrialized center in the Soviet Union. An even higher inclination orbit would have been preferred. A polar orbit was considered ideal for a recon- naissance satellite, but the Shuttle base at Vandenberg Air Force Base, California, from where such an orbit could have been safely achieved, was not operational; the weight of the larger satellites required they be launched from the Space Shuttle.

Atlantis landed safely at Edwards Air Force Base, California on March 4. No problems were reported during the brief 4-1/2 day mission, the Space Shuttle's 34th mission and the 6th mission dedicated solely to the military. (*W Post*, Mar 1/90; Mar 2/90; Mar 5/90; Feb 28/90; *NY Times*, March 1/90; Mar 2/90; Mar 6/90; *P Inq*, Mar 5/90)

• NASA Administrator Richard Truly announced the completion of the merger of the Office of Aeronautics and Space Technology and the Office of Exploration; the two offices were merged into the Office of Aeronautics, Exploration, and Technology. Arnold D. Aldrich was named to head the new NASA office, which was intended to combine the analysis of exploration mission alternatives with innovative technologies, two closely related efforts that Truly felt should proceed under a strong central management.

Truly also announced the establishment of the Office of Legislative Affairs, effective March 18, 1990. Headed by Martin P. Kress, this NASA office would be responsible for legislative matters and the coordination of budgetary and policy matters with Congress. (NASA Release 90-32; 90-33)

During February: Daniel Brandenstein, NASA's chief astronaut, returned from a tour of Soviet space facilities and said the original untended Russian shuttle, maiden voyage in 1988, would never fly again. He also reported that the Soviets planned to launch a second untended space shuttle in 1991 and automatically dock it with Mir. The station's crew planned to run some tests. Again untended, it would undock and reenter the atmosphere. (*UPI*, Feb 21/90)

• A *Chicago Tribune* article reported on the progress of commercial space launch ventures that were born out of the Challenger disaster in 1986. McDonnell Douglas and Martin Marietta made their first successful launches of satellites in 1989, and Orbital Sciences and General Dynamics were expected to follow this year. Hundreds of millions of dollars were invested by the corporations, and a backlog of missions was seen until 1995 when Japan, Russia, and China were predicted to compete with U.S. and European systems. The threat of so much competition was unnerving for the industry, but the promise of U.S. military contracts offered enough compensation for the continuation of their projects. (*C Trib*, Feb 19/90)

• The Los Alamos National Laboratory urged NASA and the National Space Council to fund a multibillion dollar nuclear propulsion project that the group believed would reduce the Mars journey by half of the expected time. The proposed system would draw on a 1970s technology, Rover and Nerva programs, that circulated liquid hydrogen through a solid reactor core, thereby heating the hydrogen and creating thrust. The reactor engine could operate only in a vacuum, so testing it on Earth would prove difficult. The program, reported *Space News*, would likely receive substantial money. (Sp News, Feb 19-25/90)

• In an unreleased report, NASA determined that the proposed untended Shuttle-C would prove too costly for use in building the Space Station. The original plan called for 18 Shuttle flights at a cost of $1.5 billion to erect the $25 billion station. Although using the Shuttle-C would reduce the number of required flights to 10 because of its more than doubled cargo space, the price tag would jump to $3.7 billion, not including the $2 billion needed to develop the vehicle. (*Sp News,* Feb 26–Mar 4/90)

• Three SR-71 Blackbird aircraft, reconnaissance planes being retired by the Air Force, were Scheduled to arrive in February at Edwards Air Force Base in California, for "flyable storage." NASA hoped to use the high-speed aircraft, capable of Mach 3 at 80,000 feet, for tests relating to the Space Plane later in the year. (*Daily News,* Jan 30/90; NASA Release 90-24)

• NASA reported on the completed computerized medical reference system, a project engendered by computer and medical experts at the University of Florida and the Kennedy Space Center in Florida in 1979 and known as the Clinical Practice Library of Medicine (CPLM). The system contains nearly all of the text and graphics of seven commonly used medical reference books that are considered essential for practicing medicine. Being lightweight as well as offering high speed access to references, the CPLM has great potential for both space travel and Earth uses alike. (NASA Release: 90-30a)

March

March 1: Telescopes for NASA's Extreme Ultraviolet Explorer spacecraft were delivered to the Goddard Space Flight Center for integration into a payload module scheduled for launch in August 1991. The science payload was designed, built, and tested at the University of California, Berkeley, to focus on the study of an unexplored portion of the electromagnetic spectrum called the extreme ultraviolet, between the x-ray and ultraviolet wavelengths. (NASA Release 90-34)

March 2: The National Research Council reported that NASA plans for accomplishing President George Bush's mandate, crew-assisted missions to the Moon and Mars, should be viewed only as a springboard to further research. Among their recommendations were greater research into nuclear power for habitable bases and space flight, research into low and artificial gravity, and greater emphasis on planning the Space Station Freedom for essential research concerning weightlessness and habitable bases. Updating Space Shuttle technology to better comply with the needs of building the Space Station was recommended as well; it was noted that the orbiter would become obsolete after the year 2000 without these changes. (*W Post,* Mar 3/90; *NY Times,* Mar 5/90; *Sp News,* Mar 12–18/90)

March 5: Japan successfully tested its new H-2 rocket, which promised to lift large, sophisticated satellites. The rocket using an LE-7 engine, similar to the high-technology propulsion system used by the Space Shuttle, was slated for service in March 1992. (*AP,* Mar 6/90)

March 6: Rockwell Space Operations was selected by NASA for a $814 million, 10-year station operation support contract. The contract would provide support for mission, flight crew, and facility operations for the Space Station and other flight programs at the Johnson Space Center in Houston, Texas. (*H Post,* Mar 7/90)

• Victor L. Peterson was named Deputy Director of the Ames Research Center in California. Since joining NASA in 1956, Petersen held several positions at the center. (NASA Release 90-36)

March 13: A House subcommittee approved a plan that authorized $1.1 billion to build the Shuttle-C. Representative Bill Nelson was chairman of the committee and a supporter of the crewless Shuttle. The Shuttle-C would carry 100,000 pounds of cargo into space; an orbiter would have only a 50,000-pound capacity. (*Sp News,* Mar 19–25/90)

March 14: A satellite launched by Martin Marietta from Cape Canaveral, Florida, failed to reach a useful orbit because of a wiring problem that did not allow separation from the Titan 3 rocket's second stage. Ground engineers successfully took the craft to a higher, but still useless, orbit, so that it would not be as likely to reenter the atmosphere. One plan offered by Orbital Sciences Corporation in Virginia was to attach a rocket to the satellite via the Space Shuttle so it could obtain the necessary 23,000 mile orbit. Martin Marietta, which had only two remaining contracts, believed it was in jeopardy of not getting any more contracts because of the incident. (*WSJ,* Mar 15/90; *W Post,* Mar 17/90; *NY Times,* Mar 16/90)

March 15: NASA's Langley Research Center selected Ball Corporation for a $90 million contract to develop an atmospheric instrument for the proposed Earth Observing System. The instrument, Spectroscopy of the Atmosphere Using Far Infrared Emission, would be designed to study middle-atmosphere ozone distribution. (NASA Release, C90-i)

March 18: The military reconnaissance satellite launched from Atlantis in February malfunctioned and was slowly spinning from orbit, said White House intelligence officials. The satellite, "vital for START (Strategic Arms Reduction Treaty) verification," was expected to reach Earth's atmosphere in April. (*W Times,* Mar 19/90; *W Post,* Mar 19/90)

March 23: Shuttle program director Robert Crippen said that a Space Shuttle rescue of the errant satellite sent up earlier this month by Martin Marietta appeared possible. He could not say whether the Intelsat communications satellite would be lifted into a higher orbit or if the Shuttle would bring it back to Earth, but noted that either approach would require about one year for planning. (*W Post,* March 24/90)

March 25: A Delta 2 rocket carried a $654 million navigation satellite, NAVS-TAR, into orbit from Cape Canaveral Air Force Station in Florida. This was the eighth launch of the Air Force's Delta 2 rocket, designed for heavier payloads than the original Delta. It was number 7 of a planned 21 satellite constellations that were to help ground units, ships, and aircraft locate their precise positions. (*UPI,* Mar 26/90)

March 29: NASA released a call for proposals to lease a commercially developed and owned space module, the Commercial Middeck Augmentation Module, which would ride in the payload bay of the Space Shuttle and add about 50 lockers to the orbiter's capacity. The need for greater capacity resulted from NASA's Commercial Development of Space Program, which was designed to enlarge research opportunities for private use by conducting experiments in materials processing, protein crystal growth, biotechnology, and fluid dynamics. (NASA Release 90-46; *Sp News,* Mar 12-18/90)

During March: A NASA study released to the Senate disclosed that designs for the planned Space Station Freedom needed significant changes that would drive up the initial cost. The study found an elaborate program of preventative maintenance that would require about 2,200 hours each year for space walks by astronauts and begin before the station was even completed. Possible solutions included use of advanced robots for maintenance, newly designed spacesuits to ease space walks, and a redesign of Space Station parts that would make them easier to repair. (*NY Times,* Mar 19/90; Mar 20/90; Mar 28/90; Mar 29/90)

• NASA's 1991 budget would be cut by $1 billion dollars, warned Congress. The Space Station, planned explorations of the Moon and Mars, and the Orbital Maneuvering Vehicle were slated to be trimmed. Money for the Earth Observation System was earmarked for an increase. (*Sp News,* Mar 26–Apr 1/90)

• The Jet Propulsion Laboratory began testing its prototype for the Planetary Rover Navigation Test Bed Vehicle, a self-navigating vehicle, about the size of a small car, for use in planned outposts on the Moon and Mars. (*Daily News,* Mar 13/90)

• Orbital Communications Corporation requested Federal approval for deploying a global satellite system for data transmission, 20 small satellites that would mostly be used for emergency services. Orbital planned to deploy them, beginning in April with its Pegasus rocket, launched from a B-52 aircraft. (*W Times,* Mar 6/90; NASA Release N90-18)

• NASA successfully tested its self-repairing flight control system on an F-15 aircraft at Ames-Dryden Flight Research Facility at Edwards Air Force Base in California. The system detected damaged flight control components, such as ailerons, rudders, and elevators, and adjusted undamaged flight surfaces to compensate so the pilot could maintain good aircraft response. (NASA Release 90-43)

• A one-year analysis of the results from a study of the ozone above the Arctic performed in the winter of 1988–1989 by Airborne Arctic Stratospheric Expedition reported a possible 17 percent loss in March. The finding was a reversal of the original conclusion that said no ozone depletion similar to that found in the Antarctic was present. (NASA Release 90-41)

April

April 2: President George Bush announced his plans for a cooperative effort with the Soviet Union in the human-assisted exploration of space, including voyages to the Moon and Mars.

On a related note, Japan agreed to open its doors to foreign-made satellites by lifting its ban on government agencies and businesses that required them to buy only Japanese spacecraft. The Japanese law was created in order to help Japan's fledgling space industry. (*NY Times*, Apr 3/90; Apr 4/90; *W Times*, Apr 4/90; *WSJ*, Apr 4/90)

• NASA and Rockwell International Corporation signed an agreement in which Rockwell would produce a cryogenic pallet aimed at extending orbiter flights from an average of 8 days to 16 days. The extended duration orbiter program involved a pallet holding spherical tanks of liquid hydrogen, liquid oxygen, valve panels, and avionics boxes for feeding additional cryogenic fluids to the orbiter's electricity-generating fuel cells. (NASA Release 90-47)

April 6: Orbital Sciences Corporation successfully launched its Pegasus rocket from a NASA B-52 aircraft to boost a 422-pound payload into polar orbit over the Pacific Ocean. The experimental rocket deployed a NASA scientific satellite and a Navy communications satellite, opening a new avenue for launching small payloads. The elimination of a costly ground booster, said company president David Thompson, provided a means by which payloads could be delivered inexpensively. (*W Post*, Apr 6/90; *WSJ*, Apr 6/90; *NY Times*, Apr 6/90; *LA Times*, Apr 6/90)

April 7: China launched a communications satellite, AsiaSat 1, from Sichuan Province for an Australian firm and became the first third-world country to enter the commercial satellite launching business. China promised to follow fair pricing practices after President George Bush lifted a sanction, brought on the country for human rights violations, that barred them from launching the U.S.-built satellite. The launch still drastically undercut U.S. and Ariane fees. (FBIS China, *China Daily*, Apr 7/90; *CSM*, Apr 9/90; *W Post*, Apr 8/90)

April 9: NASA Administrator Richard Truly named Lawrence J. Ross as Director of the Lewis Research Center (LeRc), Cleveland, Ohio. Administrator Truly announced the planned retirement of Dr. John W. Townsend and named Dr. John Klineberg to replace him as the new Director of the Goddard Space Flight Center. Sue Mathis Richards was also appointed Deputy Associate Administrator for communications. She had previously held several managerial positions in the private sector. Lawerence Ross, who joined NASA in 1963, has served mostly in an executive capacity, as Deputy Director of LeRC, as a design and test engineer, and has been given several special assignments. Klineberg joined NASA in 1970. After conducting research at the Ames Research Center, near San Francisco, California, he became Deputy Associate Administrator for Aeronautics and Space Technology and, later, Director of LeRc. Townsend began government service in 1947 and held several positions with NASA. (NASA Release 90-49; 90-50; 90-51; 90-55)

April 10: Boeing Space Operations, Inc. was awarded a $35.9 million cost-plus-award-fee contract for program assurance engineering support at Ames Research Center in California. Under the contract, Boeing would provide reliability and quality assurance, system safety engineering, test engineering, management, institutional safety, and health and environmental services. (NASA Release C90-o)

April 11: NASA unveiled its latest research laboratory at Ames Research Center in California. The Human Performance Research Laboratory was designed for study of prolonged space travel. Ground was also broken at the dedication for the new $8.6 million Automated Sciences Research Facility, designed to provide research and technology development in automation and artificial intelligence. (NASA Release 90-53)

April 16: A panel of experts convened by the National Academy of Sciences opined against a U.S./U.S.S.R. joint mission to Mars. The reason they cited was the lack of any experience in cooperation between the two and instead suggested a "graceful path of coordinated but independent explorations." (*NY Times,* Apr 17/90)

April 19: NASA scientists Michael J. Prather and Robert T. Watson warned that current levels of ozone depletion could cause public health problems and threaten plant life regardless of what steps were taken to reduce it. Their study predicted that the salient depletion over the Arctic could drift southward, similar to the damaged ozone over Antarctica that periodically drifts over New Zealand and Australia. (*P Inq,* Apr 20/90)

April 24: Space Shuttle Discovery, flight STS-31, was successfully launched from Cape Canaveral, Florida, with the $1.5 billion, 25,000-pound Hubble Space Telescope in its cargo bay. Seven years of delays, the latest occurring last April 10, when a power failure in the orbiter caused the mission to be scrubbed, preceded the event. The craft ascended to a 381-mile orbit, the highest altitude yet obtained by a Shuttle. On April 25 the Hubble telescope—which had an expected 15-year duration—was deployed; the telescope was to probe the universe with 10 times the clarity ever before achieved. Discovery landed at Edwards Air Force Base in California, on April 29, with a test of stronger, carbon brakes that would allow future landings at the Kennedy Space Center in Florida, reducing both time and money for NASA's Shuttle launches.

Scientists, however, could not receive data from the telescope immediately after it had been deployed because of an antenna entangled in a power cable. The entangled antenna could not rotate and relay information to a TDRS satellite. The low, 381-mile orbit prevented all data from reaching ground stations. The six- to eight-week chore of powering up and calibrating instruments on the telescope could not begin until the problem was resolved. Engineers zeroed in on and only partially resolved the next day. (*B Sun,* Apr 25/90; Apr 26/90; *W Post,* Apr 25/90; Apr 26/90; Apr 30/90; May 1/90; *NY Times,* Apr 25/90; Apr 26/90; Apr 30/90; *W Times,* Apr 25/90; Apr 26/90; *P Inq,* Apr 25/90; Apr 26/90)

During April: NASA began acting on some of the recommendations from its principal independent safety review organization, which had predicted that the approximately 100 flights planned by the agency would probably entail another Shuttle loss. Among the panel's main engineering concerns were the Shuttle main engine, the redesigned solid rocket motor, orbiter structural loads, long-term orbiter maintenance, and human factors. NASA Associate Administrator William Lenoir explained that the safety panel acts as a "devil's advocate" and noted that conclusions drawn by them were intentionally exaggerated. (*AvWk,* Apr 23/90; *CSM,* May 8/90)

• NASA tested two of its experimental aircraft at the Ames-Dryden Flight Research Facility at Edwards Air Force Base in California. One of the aircraft, an F-16XL, used an experimental wing surface for improving airflow at supersonic speeds. The second aircraft, an X-29, demonstrated an increase for high-angle-of-attack, the angle of an aircraft relative to its actual flight path. (NASA Release 90-52; 90-59)

• Two NASA sounding rockets were scheduled for launch from the White Sands Missile Range in New Mexico to observe Comet Austin. The two-stage, suborbital Black Brant IX rockets were set to carry a faint object telescope and a spectrograph (for observation in the far ultraviolet spectral range) for Johns Hopkins University and a far ultraviolet spectrometer for the University of Colorado. (NASA Release 90-57)

May

May 1: NASA's Goddard Space Flight Center, Greenbelt, Maryland, awarded Jackson and Tull of Washington, D.C. a $19 million contract to support the Applied Engineering Division in automation and robotics and thermal and data systems. The firm was selected under Section 8(a) of the Small Business Act; the potential for the contract over a five-year period was $63 million. (NASA Release C90-q)

May 2: NASA named William W.L. Taylor of TRW Corporation as the Chief Scientist for the Space Station Freedom Program. Taylor, who was with NASA until 1978 when he joined TRW, was the third person chosen for the two-year assignment and the first from industry. (NASA Release 90-61)

May 3: NASA was scheduled to launch two Multiple Access Communications Satellites (MACSATs) later in May aboard a Navy-owned Scout rocket. The MACSATs were designed for a global store-and-forward message relay. (NASA Release 90-64)

May 4: Engineers at NASA ran tests of the Hubble Space Telescope to determine if vibrations were interfering with precise pointing at specific stars. The problem had been discovered two days earlier when sensors locked on to a "guide star" to orient the spacecraft. Engineers noticed a slight wobble, one six-hundredth of a degree up and down every minute, that interfered with both fine pointing mechanisms and star observations. (*NY Times,* May 5/90)

May 8: The Hubble Telescope failed a test on this date as scientists attempted to direct it toward stars emanating a common brightness; the telescope was able to find none. It was later discovered that the problem originated from an error in the telescopes computer guidance system, and a new program with an updated star chart was sent up. (*NY Times,* May 10/90; *W Post,* May 10/90)

May 14: Jerry Grey, Director of Science and Technology Policy for the American Institute of Aeronautics and Astronautics, recommended that weapons designed for the Strategic Defense Initiative be developed to locate and destroy any asteroids that threaten Earth, views endorsed by Vice President Dan Quayle. Grey noted that 1,500 asteroids cross the Earth's orbit every year, that the location of only 100 of them is known, and that the impact of even a relatively small asteroid could have a devastating effect. (*P Inq,* May 15/90; *C Trib,* May 17/90)

• Lockheed Corporation received a $971 million contract from NASA to oversee design and development of a new generation of rocket motors for the Space Shuttle. Twenty more powerful solid rocket motors were to begin lifting the Shuttle in 1995, according to the contract, and sources said that if NASA exercised the option to order 88 more, the contract would be worth $1.39 billion. (*WSJ*, May 15/90; NASA Release 90-68)

May 15: NASA said that it would rescue the uninsured communications satellite that Martin Marietta Corporation boosted into a useless orbit last March 14. The cost was set at $1 million to either retrieve it or attach rockets, that would send it into a higher orbit, in 1991 or 1992. Intelsat said that it would consider the cost at its next board of governors meeting. (*W Times*, May 16/90)

May 20: Ground controllers adjusted commands on the Hubble Space Telescope for focusing on stars. Although other problems existed—wobble from solar panels on the telescope and vibration as it went through space—the first pictures were taken and sent to Earth on this day. (*NY Times*, May 21/90; *W Times*, May 21/90; *W Post*, May 21/90)

May 24: The Department of Defense and NASA announced the appointment of contractors to continue research and development on the National AeroSpace Plane. General Dynamics, McDonnell Douglas, United Technologies (Pratt & Whitney), and Rockwell International planned to jointly pursue hypersonic technologies. (NASA Release 90-71)

May 31: The Apollo 204 Capsule, which caught fire on the launch pad at Cape Canaveral, Florida, and claimed the lives of three astronauts on January 27, 1967, remains at Langley Research Center in Virginia, where it has been stored since shortly after the accident. The capsule was slated to be entombed with the ill-fated Space Shuttle Challenger in an abandoned missile silo at Cape Canaveral, Florida. (NASA Release 90-76)

June

June 1: NASA launched a Delta rocket carrying a West German observatory into a 360-mile-high orbit from Cape Canaveral, Florida. The observatory consisted of an x-ray telescope and a wide-field camera for detecting extreme ultraviolet light; scientists hoped the observatory would yield information about forces associated with the center of our galaxy, intergalactic gases, quasars, black holes, hot stars, and objects on the edge of the universe. The observatory was dubbed Rosat, after the German Scientist Wilhelm Roentgen who discovered x-rays in 1895, and promised to be 100 times more sensitive than even the Einstein x-ray Observatory launched in 1978. It was originally scheduled for launch aboard a Space Shuttle in 1987. (*LA Times,* Jun 2/90; *NY Times,* Jun 2/90; *B Sun,* Jun 2/90; *P Inq,* Jun 2/90)

June 5: A letter from Lennard A. Fisk to Jacques Breton of the French space agency, the Centre National d'Etudes Spatiales, revealed NASA's defunct plans for a U.S.-French experiment that was to fly aboard a 1994 Soviet mission to measure the mineral content of Mars. The constraints recently imposed on the two countries was the reason given for cancellation of the project. (*W Post,* Jun 18/90)

June 6: A fuel leak that postponed Space Shuttle Columbia's May 30 launch was located, and engineers decided the orbiter would have to return to the hanger for repairs. NASA officials were not sure when the craft would be launched but announced that only eight of the nine planned flights for the year would be flown. (*NY Times,* Jun 7/90; *W Times,* Jun 8/90)

June 7: NASA canceled a contract with TRW Corporation for the Orbital Maneuvering Vehicle, also known as the Space Tug, a spacecraft intended to lift satellites (specifically the Hubble Space Telescope and a proposed X-ray telescope) into higher orbit. The reason given for the cancellation was recent budget pressure facing NASA. (*WSJ,* Jun 8/90; *W Times,* Jun 8/90; *Sp News,* Jun 11–17/90; NASA Release 90-78)

• NASA announced that Steven A Hawley had been named Associate Director at Ames Research Center, California. He was selected as an astronaut in 1978 and was the veteran of three Shuttle flights. (NASA Release 90-77)

June 8: The Air Force launched a Titan 4 rocket from Cape Canaveral in Florida and put a classified payload into orbit. The Titan 4, manufactured by Martin Marietta Corporation, was designed to replace the Space Shuttle as the military's principal means of reaching space. This was the second successful launch of the rocket, although the maiden voyage in 1989 almost resulted in a catastrophe because of a malfunctioning engine nozzle. (*W Post*, Jun 9/90; *NY Times*, Jun 9/90)

June 13: NASA said it had received the go ahead from Intelsat to rescue the Intelsat 6 satellite put into a low-orbit three months earlier. Plans called for lifting the satellite into a higher orbit sometime in 1992; sources said that the mission would probably be carried out during the inaugural flight of Shuttle Endeavor. (*B Sun*, Jun 14/90; *W Times*, Jun 14/90)

June 14: NASA engineers revealed that problems with the Hubble Space Telescope were even more difficult to correct than first expected. The most vexing problem was a 20-minute vibration from the solar panels every 90 minutes as the craft passed from night to day. The onboard computer's attempt to compensate for the vibration only made the problem worse. Another problem was the telescope's frequent inability to lock onto the stars that it was commanded to seek. In addition, the reaction of the spacecraft to radiation over the South Atlantic, where the Van Allen belt dips closer than usual to the Earth, was more acute than expected; numerous data were lost and false computer signals were given. It was hoped that all three problems could be remedied via modifications to the telescope's computer. (*NY Times*, Jun 15/90; *P Inq*, Jun 15/90; *B Sun*, Jun 15/90)

• NASA's Office of Commercial Programs sponsored plans to develop a system for launching and recovering commercial spaceborne experiments. Assisted by numerous universities, the Commercial Experiment Transporter would be launched on an expendable rocket and contain both a recovery system and a service module. The two would then separate so that the experiments aboard the recovery system could land in the United States while those not requiring recovery could remain in orbit. (NASA Release 90-83)

June 21: Contel Federal Systems transferred title of the Tracking and Data Relay Satellite System to NASA, effective July 1, 1990. Contel transferred the system 42 months earlier than called for, saving the Government $16 million. (NASA Release 90-86)

• The Ames Research Center in California awarded a five-year, $32.8 million contract to Silicon Graphics for a work station to be used in the Ames Numerical Aerodynamic Simulation Processing System Network. The work station would allow users to analyze and visualize their work in computational fluid dynamics and other large-scale scientific simulation and modeling applications. (NASA Release C90-w)

June 23: Martin Marietta Corporation successfully launched a Titan 3 rocket from Cape Canaveral, Florida, on this date that carried a replacement satellite into orbit for the Intelsat 6 stranded in March. The new communications satellite was to join 14 others in September and would transmit 3 television channels and 120,000 telephone calls simultaneously. (*P Inq,* Jun 24/90; *W Post,* Jun 24/90)

June 27: NASA said that the Hubble Space Telescope, which had been plagued with problems, was discovered to have a flawed mirror, causing the $1.5 billion instrument to be "near sighted." The mirrors, which were ground into the wrong shape, rendered useless the telescope's most important scientific instrument, a widefield planetary camera designed to explore the farthest reaches of the universe. Engineers said that instruments scheduled to be attached to the telescope in 1993, 1996, and 1997 could compensate for the flaw and added that these flight schedules might be advanced. Tests were not done on the effectiveness of the mirrors before the telescope was launched because engineers felt that the checks were cost prohibitive. (*P Inq,* Jun 28/90; *NY Times,* June 28/90; *W Post,* Jun 28/90)

June 28: Ames Research Center in California awarded a cost-plus-award-fee $210.8 million contract for software support. Under the contract Sterling Federal Systems, Inc. would provide computational support for problem applications and systems programming, systems design and engineering, software management and maintenance, and modification of existing software. (NASA Release C90-y)

June 29: A hydrogen leak similar to the one that scrubbed Columbia's mission late in May was discovered in the Space Shuttle Atlantis as it was being readied on the launch pad. Engineers determined that Atlantis would need to return to the hanger where it could be separated from its external fuel tank for repairs. Officials said that the flight, originally scheduled for mid-July, would be delayed at least several weeks and noted that the entire fleet was grounded until the problem could be ascertained. Finding the cause of the leak had proven difficult. If the final analysis pointed to a design flaw, grounding might have been stretched to months or even years. (NASA Release 90-89; *W Post,* Jun 30/90; *NY Times,* Jun 30/90; *WSJ,* July 2/90; *CSM,* Jul 2/90)

During June: NASA awarded a $1.3 million contract to Martin Marietta Corporation to study design concepts for a space based "service station." The system would service satellites beyond the range of the Space Shuttle, provide them with fuel as well as replace certain components. (*W Times,* Jun 7/90)

• The Air Force and NASA entered their final stage of a $250 million, ten-year experiment for the X-29 program, scheduled to be completed in March 1991. The test plane was designed to be 30 percent unstable and was dependent upon a sophisticated flight control system to produce and maintain stability. Glenn Spacht of Grumman Corporation, developer of the X-29 and producer of the F-14, said,

"When applied to the F-14, the plane will get better handling qualities, survivability, maneuverability, and antispin characteristics. . . .The F-14 can certainly reap the benefits of the X-29." (*Def News*, Jun 25/90)

• Scientists and engineers at the Jet Propulsion Laboratory in Pasadena, California began extensive tests on its recently developed untended planetary rover. An unproven technology was needed for a vehicle on a distant planet controlled by ground crews on Earth inasmuch as round-trip communication would be anywhere from 8 minutes to 40 minutes. Computer Aided Remote Driving allowed vehicle planning and identification of a 10s-of-meters-long-obstacle-free path, with a three-dimensional display from stereo cameras aboard the vehicle. Semi-Autonomous Navigation allowed an operator to determine a nominal extended route tens of kilometers long. (NASA Release 90-85)

July

July 2: NASA's Marshall Space Flight Center in Huntsville, Alabama, awarded a $42.6 million contract to Rockwell International for continued Space Shuttle systems engineering and integration services. The contract called for Shuttle launch and flight support, test verification, operations integration, flight evaluation, engineering analysis, and data systems operations, as well as special studies. (NASA Release C90-z)

July 3: NASA and Orbital Sciences Corporation signed a five-year agreement in support of the firm's Pegasus and Taurus launch vehicle programs. As part of a continuing effort to foster a strong U.S. commercial launch industry, Orbital Sciences was given access to agency launch support property and services. Prior to this agreement, similar agreements were made with General Dynamics, McDonnell Douglas, Martin Marietta, and the LTV Corporation. (NASA Release 90-92)

July 5: The Hubble Space Telescope Optical Systems Board of Investigation created by NASA met for the first time in Washington, D.C., to hear presentations from the Marshall Space Flight Center in Alabama, Huntsville, and the Hughes Danbury Optical Systems, makers of the precision mirrors for the telescope. Members of the Board included Lennard A. Fisk, Associate Administrator for Space Science and Applications; Dr. Lew Allen, Director, Jet Propulsion Laboratory; and Gary Tesch, Board Counsel. At the conclusion of the meeting, Dr. Allen said, "The board now has a good base of information from which to proceed with its investigation." (NASA Anno)

July 8: President George Bush said he would allow U.S. commercial satellites to be launched by Soviet rockets. The rockets would be fired from a projected site in Australia, an advantageous position for orbits, located just 12 degrees south of the equator. (FBIS Sov, Jul 9/90; *NY Times,* Jul 9/90)

July 9: Two Space Shuttle crew commanders were grounded by NASA for disciplinary reasons. Robert L. Gibson was grounded for one year after he flew a stunt plane, in a July 7 Texas air show, that collided with another plane and killed the second pilot. David M. Walker was grounded for 60 days because the T-38 jet aircraft he piloted in May 1989 nearly collided with a Pan Am Airbus 310 over Dulles Airport, according to the second pilot. (*P Inq,* Jul 10/90; *W Times,* Jul 10/90; *NY Times,* Jul 10/90; NASA Release 90-96)

July 10: NASA's Chief Scientist William Lenoir told a Senate committee that the Agency should have performed tests on the Hubble Space Telescope's mirrors. He said that the problem could have been caught by simpler, less costly tests than the "end-to-end" test, rejected because of its high cost. (Hearing before the Subcommittee on Science, Technology and Space, Jul 10/90; *B Sun,* Jul 11/90; *W Post,* Jul 11/90; *NY Times,* Jul 11/90)

July 13: Engineers located two leaks in the Space Shuttle's fuel line that they determined could easily be fixed and stated that flights would resume soon. The problems, said William Lenoir, NASA's flight director, did not appear to be a result of any design flaw; he added that eight of the nine scheduled missions would likely be flown this year.

In the meantime, investigators from a six-member board of investigation created by NASA July 2 were seeking to find what had gone wrong with the Hubble Space Telescope. Lew Allen, Director, Jet Propulsion Laboratory, who headed the board, promised a congressional committee that they would pinpoint the problem and fix responsibility. The investigation, he said, would be "very embarrassing to someone or some group." (*W Post,* Jul 14/90; *NY Times,* Jul 14/90)

July 15: Recent grounding of the Space Shuttle fleet and problems with the Hubble Space Telescope had prompted President George Bush to consider appointing a panel composed of a group of experts outside of Government to review the space program, White House officials announced. Officials noted, however, that NASA made a good call when it grounded the Shuttle fleet and that no plans for changing its management had been formulated. Vice President Dan Quayle met with NASA Administrator Richard R. Truly on July and directed him to appoint the independent task force. (*NY Times,* Jul 16/86; Jul 17/90; *B Sun,* Jul 16/86; Jul 17/90; NASA Anno)

July 17: NASA scientists proposed almost a dozen plans for a temporary fix of the Hubble Space Telescope until 1993 Space Shuttle mission would install an optically corrected main camera. A brainstorming session included some "far fetched" ideas that involved Shuttle missions to the telescope.

Elsewhere, NASA official William Lenoir reported that Space Shuttle flights might resume soon and that Atlantis could fly as early as August 10. He added that if the repairs could not be accomplished on the launch pad, then the orbiter would be taken into the hanger and rescheduled for launch sometime after the October 5 liftoff of Discovery. (*B Sun,* Jul 18/90; *W Post,* Jul 18/90; *NY Times,* Jul 18/90)

July 20: A report released by NASA, concerning plans for the Space Station Freedom, said that 3,700 hours of maintenance would be required annually to keep it running. The final figure from the six-month study was up from the

2,200 hours that a preliminary version had given in March and dwarfed the original goal of 130 hours. The most alarming statistic was that 6,200 hours of maintenance would be needed before the station could even be habitable. "The significance is that there's no one up there to do this," said Dr. William J. Fisher, an astronaut who headed the 12-member panel that did the estimate. (*NY Times*, Jul 21/90; *W Post*, Jul 21/90; *WSJ*, Jul 23/90)

July 24: A NASA official announced that Mark Showalter, at the Ames Research Center in California, had discovered another moon orbiting Jupiter. He found it while doing analysis of images taken in 1980–1981 by Voyager 2. The smallest of the planet's known satellites was dubbed 1981S13. (*B Sun*, Jul 25/90; *W Post*, Jul 25/90; NASA Release 90-103)

• An Ariane rocket carrying both a French and a German TV satellite was successfully launched from French Guiana. The rocket was identical to the one that exploded shortly after liftoff last February 22. (*W Post*, Jul 25/90; *WSJ*, Jul 25/90)

July 25: The hydrogen leak that was detected in the Space Shuttle Atlantis reappeared, and engineers determined that the Shuttle would have to be taken off the launch pad and rolled back to the hanger for a closer look. Columbia, undergoing repairs for a similar leak, was expected to be on the launch pad early in August. The Atlantis liftoff with its classified military payload were rescheduled for November, bumping another launch into 1991 and reducing the projected number of flights in 1990 to seven. (*P Inq*, Jul 26/90; *NY Times*, Jul 26/90; *W Times*, Jul 26/90)

• After three delays, NASA and General Dynamics Corporation launched a Delta rocket from Cape Canaveral, Florida, that carried a $189 million satellite to study the Earth's magnetic and electrical fields. The Combined Release and Radiation Effects Satellite (CRRES), a joint effort by NASA and the Department of Defense, carried 24 canisters containing chemicals that were to be released in fall 1990 and become ionized by the sun's rays. Visible clouds would then spread along magnetic field lines, allowing scientists to see how electrical fields interact with normally invisible charged particles. The CRRES releases were to be augmented by chemicals released from 10 sounding rockets launched from Puerto Rico and the Marshall Islands. (*P Inq*, Jul 26/90; *NY Times*, Jul 26/90; NASA Release 90-97; 90-98)

• The United States and the Soviet Union signed an agreement in which NASA's Total Ozone Mapping Spectrometer (TOMS) would fly on a Soviet Meteor-3 spacecraft sometime in 1991. A TOMS instrument had monitored ozone concentrations from a Nimbus-7 satellite since 1978, focusing on the southern hemisphere and the development of the Antarctic ozone hole. This satellite, however, had already performed well beyond its designed lifetime, and the TOMS/Meteor-3 mission was slated to replace it. Another TOMS satellite was scheduled to be aboard a U.S. spacecraft in 1993; two or three years subsequent to that mission another TOMS was scheduled for a Japanese ADEOS satellite. (NASA Release 90-105)

July 26: NASA and Japan's Ministry of State for Science and Technology agreed on new areas of space cooperation during a meeting in Tokyo. NASA/Japan projects agreed upon were observation of the ozone layer from satellites, space environment monitoring (real-time solar wind data), Space Station solar terrestrial physics carried out from the Space Station Freedom, and space microgravity experiments aboard the Space Shuttle. Other areas discussed were the ongoing cooperation for building and utilization of the Space Station and joint projects in x-ray and infrared astronomy, cosmic ray research, the study of ocean dynamics, measurement of cloud height by satellite stereograph, reception of Japan's first marine observation satellite data, and satellite measurements of tropical rainfall. (NASA Release 90-106)

July 31: State officials said that the Soviet Union had requested use of facilities at Cape Canaveral, Florida, to launch their Proton rockets. The request followed their stated intentions of joining the commercial space race. University of Houston professor Peter Bishop explained that a launch pad closer to the equator meant less fuel would be needed for achieving proper orbit. (*B Sun,* Aug 1/86)

During July: NASA and U.S. university scientists and Canadian researchers joined in a study of pollution at high northern latitudes resulting from the release of methane, an important greenhouse gas, from tundra, forests, and marshes. Both ground and air measurements were taken; NASA's main sampling platform was a Lockheed Electra aircraft. Research began July 5 and was scheduled to conclude on August 20, 1990. (NASA Release 90-102)

• The Goddard Space Flight Center in Greenbelt, Maryland, selected Stanford Telecom for negotiations leading to a five-year, $30 million contract to provide systems engineering support for the Advanced Tracking and Delta Relay Satellite and the Advanced Network Control Center. Also, Martin Marietta Corporation received a $326.8 million contract from the Air Force for modifications to space launch facilities at Cape Canaveral, Florida. (NASA Release C90-aa)

• Scientists at the Jet Propulsion Laboratory in Pasadena, California, devised a way to salvage the Japanese lunar mission that successfully orbited a probe around the Moon but did not send data back because of a failed transmitter. The mother ship that launched the probe was in an elliptical orbit around the Earth and could achieve lunar orbit by taking advantage of the "fuzzy boundary". The "fuzzy boundary" is an area about one million miles from Earth where gravitational forces emanating from our planet and the Sun cancel each other out. The limited supply of fuel carried by the spacecraft was not sufficient to allow it to reach a lunar orbit by conventional means. Success with the plan, they determined, would mean an enormous fuel savings should lunar bases be built in the future. *(LA Times,* July 16/90)

August

August 2: The Air Force launched a Delta rocket from Cape Canaveral, Florida, which successfully carried a NAVSTAR satellite into orbit. The navigation satellite was the eighth in a planned series of 21 that promised to give military units their locations to within 50 feet. (*W Post,* Aug 3/90)

• Officials announced that an advisory committee created by Vice President Dan Quayle and the National Space Council was set up to examine how to manage future efforts in space. The committee, made up of two former congressmen, a former astronaut, and several top industry executives, was to report to the National Space Council in about four months. Quayle said that he had urged the committee's creation because of problems with the Hubble Space Telescope, grounding of the Shuttle fleet, and design flaws in the proposed Space Station. (*NY Times,* Aug 3/90)

August 8: Government officials said that the mirrors attached to the Nation's next generation of weather satellites used to reflect images from Earth's atmosphere into internal sensors would warp from the Sun's heat, according to recent tests. NASA formed a team of nine experts from Ford and ITT, makers of the spacecraft, to correct the problem in time to meet the 1992 launch date. A single Geostationary Operational Environmental Satellite was in orbit at the time. Officials were worried that, if it failed, the Nation would have to rely on planes, radar units, and other satellites that were not capable of giving the comprehensive overview needed, for example, to follow storms. There were normally two satellites of this type in orbit, but one was destroyed in 1986 when a Delta rocket malfunctioned. Solutions for the GOES-NEXT flawed mirrors ranged from repairing them, which would cause a six-month delay, to re-manufacturing, and causing a delay of more than one year. (*NY Times,* Aug 8/90; *W Post,* Aug 9/90; *W Times,* Aug 9/90)

August 10: The $744 million Magellan spacecraft launched in May 1989 reached Venus and achieved polar orbit for a mission to map the entire planet. The radar-produced images promised to be 10 times more detailed than those from previous expeditions that had mapped only sections of Venus. Information from the mission could also provide a better understanding of predicted global warming, or the greenhouse effect, said NASA scientist Gentry Lee. (*NY Times,* Aug 11/90; *W Post,* Aug 11/90; *LA Times,* Aug 11/90; *C Trib,* Aug 11/90)

August 13: Surprised NASA scientists released a photograph taken August 3 by the Hubble Space Telescope that provided unexpected detail of a young star system known as 30 Doradus, 160,000 light years from Earth. After scientists corrected the flawed mirror problem with computer enhancement, the telescope revealed 60 of what were yet the youngest and heaviest stars known. The "star nursery," was once thought to have only one large star; the most advanced telescope prior to the Hubble telescope had raised that number to 27. (*W Post,* Aug 14/90; *W Times,* Aug 14/90; *P Inq,* Aug 14/90; NASA Release 90-111)

August 16: Charles Pellerin, director of astrophysics for NASA, announced that problems with the Hubble Space Telescope mirrors could easily be fixed in a Space Shuttle mission scheduled for mid-1993. The mission would require replacing the telescope's widefield planetary camera with one designed to correct the flawed mirrors. (*P Inq,* Aug 17/90; *W Times,* Aug 17/90)

August 17: NASA scientists regained contact with Magellan after a 14-hour loss. The spacecraft went into a safe mode, in which it shut down many of its systems and pointed its solar panels toward the Sun because of navigation problems it falsely interpreted. The ship's computer memory was sent back to Earth so scientists could determine the problem in Magellan's program. Scientists were pleased, however, with the first set of pictures taken during Magellan's test run over the planet. They, however, cautioned that Magellan's mission to begin regular mapping by August 29 would likely be delayed. (*W Times,* Aug 20/90; *P Inq,* Aug 19/90; *W Post,* Aug 18/90; *NY Times,* Aug 18/90; *LA Times,* Aug 18/90)

August 18: A Delta rocket carried a British television broadcasting satellite, Marcopolo 2, and put it into orbit from Cape Canaveral, Florida. The only problem for the ninth U.S. commercial launch of a payload into orbit was a two-hour delay caused by thunderstorms in the area. (*AP,* Aug 18/90)

August 20: NASA Administrator Richard R. Truly announced the establishment of a Space Commerce Steering Group, composed of NASA officials, that would provide an overview of commercial space technology applications. The goal of the Group was to provide coordination of potential commercial capabilities; NASA's Office of Commercial Programs would remain the primary action agent. Truly maintained that NASA "must ensure that [its] technology is transferred to the private sector. The taxpayer's investment in NASA is an investment in the international competitiveness of U.S. industry." (NASA Release 90-112)

August 21: Scientists at NASA said that the Magellan spacecraft seemed to be working. Engineers had received all of the data from the craft's computer memory and were in the process of loading new programming. Officials said the navigation problem might have been caused by a series of rare events; they hoped it would be fully operational soon. (*NY Times,* Aug 22/90; *W Post,* Aug 22/90; *W Post,* Aug 22/90)

• A report from a panel convened by the National Research Council and released to the White House recommended smaller and simpler satellites to monitor the Earth's climate, as opposed to the $30 billion, six large satellites being planned by NASA. The expense of the Earth Observing System (EOS) could take funds away from other important studies and future budgets. It could result in a delay in launching the EOS spacecraft, the report warned. Noting that at least one large platform was necessary for a number of scientific instruments to be grouped together, the panel recommended that three of the large satellites be replaced by smaller ones. (*NY Times,* Aug 22/90; *W Post,* Aug 22/90)

August 27: NASA responded to a request issued August 7 by the Office of Management and Budget asking for various Federal agencies to outline the effects of a $100 billion cut in the 1991 budget. The report submitted by NASA warned of a halt in Space Shuttle flights, suspended development of the Space Station, and delayed launch of the Mars Observer spacecraft. Although not expected, cuts across-the-board were a possible result of the Gramm-Rudman deficit control law. (*Sp News,* Sep 9-13/90)

August 29: The first scientific results from the Hubble Space Telescope confirmed that it would reveal a more detailed view of the heavens than any ground tele- scope, in spite of its flawed mirror. The latest pictures showed the clearest view yet of a supernova discovered in 1987 and of densely packed stars in a distant galaxy that could be the result of a black hole.

Meanwhile, scientists at NASA reported that the Magellan spacecraft was operating in another safe mode and communication with it had been off-and- on. The navigation problem had been narrowed down but not pinpointed, and the mapping survey of Venus would be delayed two or three weeks. (*NY Times,* Aug 30/90; *W Times,* Aug 30/90; NASA Release 90-117; 90-118)

During August: Joseph B. Mahon, Deputy Associate Administrator, and astro- naut Michael J. McCulley announced they would retire. Mahon had served with NASA for almost 30 years, managing and directing untended launch vehicles, Shuttle carrier systems, and spacelab operations. Captain McCulley, who planned to retire from the Navy, was to join Lockheed Space Operations Company and be involved with day-to-day processing of Space Shuttle vehicles. (NASA Release 90-116; 90-119)

• NASA announced the selection of Margaret G. Finarelli as Acting Associate Administrator for External Relations, and the names of microgravity mission payload specialist candidates. Finarelli, who had been Deputy Associate Administrator since December 1988, would temporarily replace Kenneth S. Pedersen, who had taken a one-year teaching assignment with Georgetown University in Washington, D.C. Candidates for payload specialists for the

March 1992 STS-53 Space Shuttle mission were Lawrence J. DeLucas, Joseph Prahl, Albert Sacco, and Eugene H. Trinh. (NASA Release 90-107; 90-108)

• A Government industry board selected the gas generator power cycle engine, similar to the J-2 engine used on the Saturn moon rockets, for the rocket engine to be designed and built to power the NASA/USAF Advanced Launch System. Chosen over the closed expander power cycle, the engine would power next century launch vehicles, capable of delivering a wide range of payloads into Earth orbit at a reduced cost. (NASA Release 90-113)

• In a flight test that concluded in August, NASA, the Air Force, and Boeing Commercial Airline Group announced better-than-expected findings taken from the results of modifying a 22-foot section of a Boeing 757 wing. The design reduced drag in what could equate to 10 percent if the entire span of both wings had been modified. A one-percent reduction would have amounted to a $100 million annual savings in fuel costs for the U.S. airline industry. (NASA Release 90-115)

September

September 5: Liftoff of the Space Shuttle Columbia was scrubbed for the third time, after a fuel leak was discovered. It was reportedly not in the same area as the leak that halted the mission last May. The second delay for liftoff occurred during the last week of August because of communication problems with the x-ray telescope in Columbia's cargo bay. NASA engineers believed the problem lay in the fuel pumps feeding the Shuttle's three main engines and began replacing them the next day. Agency officials said that the flight would be delayed at least 11 days. (*W Times,* Sep 6/90; Sep 7/90; *W Post,* Sep 6/90; Sep 7/90; *NY Times,* Sep 6/90; Sep 7/90)

September 7: A Titan 4 rocket booster exploded at Edwards Air Force Base in California during tests, killing one worker and injuring nine others. The Air Force was further developing the rocket, which experts called problem plagued, hoping to replace the Space Shuttle for launching of heavy payloads. (*C Trib,* Sep 8/90; *LA Times,* Sep 9/90)

September 13: NASA Administrator Richard Truly told a panel created by President George Bush that the Agency was facing serious problems, but not what most people believed them to be. The two obstacles standing in front of the Agency, he said, were "Governmentwide practices" that encumbered connections between programs and resources and the public's misunderstanding of the proposed Space Station Freedom. The Space Station's design flaws, he added, were either solved or close to a solution, "yet public understanding is to the contrary."

In a related matter, *Federal Computer Week* reported that NASA planned to spend $1.9 billion over the next five years on its seven centers of excellence in the area of high-performance computing. The Centers would focus on training college- and graduate-level students, thereby educating the next generation of scientists in the field. Truly, who spoke at a conference on supercomputing in September noted, "NASA needs high-performance computing. Some of our missions depend on it." (*W Post,* Sep 14/90; *Fed CoWk,* Sep 10/90)

September 15: Magellan officially began its mission of mapping Venus on this date when it used radar to photograph the planet for the first time since radio contact was lost August 16. Computerized commands were radioed to the spacecraft, and confirmation that it had begun its 243-day chore was returned 13 minutes later. Pictures revealing mountains, valleys, and a unique looking meteor crater were received by scientists two days later. (*P Inq,* Sep 16/90; Sep 18/90; *W Post,* Sep 25/90)

September 17: Liftoff of Columbia was again scrubbed because of another hydrogen leak. This time the suspect was gritty sandpaper residues from a mobile launch platform restoration. A postponement period for the launch was indefinite, but officials said that it would be delayed until after the October 5 Discovery flight. (*W Times,* Sep 18/90; *W Post,* Sep 18/90; *NY Times,* Sep 18/90)

September 18: The House voted to slash the Strategic Defense Initiative $4.7 billion request to $2.3 billion. Representative Jon Kyl argued that the United Stated needed "Star Wars" to avoid "being held hostage by some tinhorn dictator." Representative Ronald Dellums retorted that only $143 million of the program's budget was dedicated to destroying short-range missiles, those developed by Third World Nations. (*P Inq,* Sep 19/90)

September 19: The House Science, Space, and Technology Committee recommended an increase of NASA's budget from $12.3 billion to $17 billion for the next year, including the $2.9 billion earmarked for continued development of the Space Station. Committee members acknowledged that the final spending plan would probably be lower but wanted to show their support despite NASA's recent problems. (*P Inq,* Sep 20/90)

September 20: NASA took the Space Shuttle Columbia off its schedule for 1990 launches, operating under the assumption that the orbiter was prone to fuel leaks. Still unable to pinpoint the exact location of the last hydrogen leak, NASA hoped that the craft would be able to fly in early 1991. A team of nine engineers was working on the problem. (*NY Times,* Sep 21/90; *W Post,* Sep 20/90; *C Trib,* Sep 20/90; NASA Release 90-127)

September 27: The Federal Aviation Administration signed a memoranda of understanding with NASA to jointly research problems in aviation. Areas to be investigated were environmental compatibility, human factors, severe weather, integration of cockpit and air traffic operations, airworthiness, and noise. (NASA Release 90-131)

September 28: The House passed a $44 billion, three-year spending authorization bill for NASA that included establishment of a Moon base and a crew-assisted expedition to Mars. The bill was to next go to a conference committee to work out differences with the Senate. (*LA Times,* Sep 29/90)

During September: Astronaut Bruce McCandless retired from NASA and as Captain from the U.S. Navy. He helped develop the space agency's Manned Maneuvering Unit, a jet-powered backpack he was the first to wear during a Space Shuttle Challenger mission in February 1984. (*W Times,* Sep 12/90)

• The Langley Research Center in Virginia selected Mason and Hanger Services, Inc. for negotiation of a five-year, $97.9 million contract to support all aspects of supply, transportation and equipment management; general and specialized administrative services; different aspects of the security program; and support of the scientific and technical information program.

The Television Development Division of NASA Headquarters awarded the Imaging Systems Laboratory of Florida Atlantic University a one- to five-year contract with a potential value of $3 million for research and development in advanced video systems for space and terrestrial applications. (NASA Release C90-cc; 90-131A)

• A joint NASA/U.S. Geological Survey team of American scientists planned to work with a Soviet Science team to study volcanoes along Russia's Kamchatka Peninsula, part of the Pacific's "Ring of Fire". The study would mark the first time Western scientists had been allowed in the region. (NASA Release 90-124)

• The Air Force decided in September to convert the Space Launch Complex-6 at Vandenberg Air Force Base in California for Space Shuttle liftoffs into a launch pad for its Titan 4 rockets. The launch pad had been idle since the Challenger accident in 1986. (*NY Times*, Sep 30/90)

October

October 1: The High Resolution Imager, an instrument for the study of cosmic x-rays aboard the German Roentgen Satellite launched last June, sent back its first light pictures to a ground station near Munich. The cooperative program between the United States, West Germany, and Great Britain, called ROSAT, excited several international scientists. The pictures sent back represented a neutron star 3,000 light years from Earth, a supernova that occurred in this galaxy 320 years ago and 10,000 light years away, and a cluster of galaxies known as Abell 2156. (NASA Release 90-133)

October 3: The Senate approved $13.4 billion for NASA's 1991 budget. The bill drastically reduced the Moon and Mars initiative and cut $863.6 million from the planned Space Station. The measure was to go to a conference committee to iron out differences with the House appropriations bill before being sent to President George Bush. (*WSJ,* Oct 4/90)

October 4: NASA scientists reported that despite its flaws, the Hubble Space Telescope was sending back some clear pictures, better than any a ground telescope could produce. The latest photographs of Jupiter and Pluto were better than any ever taken, but until the telescope's "near sightedness" was corrected images of distant, faint objects would continue to be unclear. (*W Post,* Oct 5/90; *NY Times,* Oct 5/90)

• The official Soviet Tass News Agency reported that their most advanced rocket, a Zenit booster, had blown up seconds after liftoff and destroyed much of the launch pad in the Kazakhstan Republic. The payload, said Western analysts, was likely a recon- naissance satellite. The rocket had earlier been promoted for sale in the West and had been used to fly some versions of the Soviet Space Shuttle. (*W Post,* Oct 12/90)

October 6: Space Shuttle Discovery, flight STS-41, was launched from Cape Canaveral, Florida after a five-month dry spell; the fleet was grounded during much of this period because of hydrogen leaks. Ulysses, the craft's 814-pound nuclear powered payload, was released later that afternoon, and a three-stage solid-fuel rocket began the craft on its long trip to Jupiter. Ulysses was to sling shot Jupiter in order to achieve polar orbit about 120 million miles from the sun in May 1994. In order to escape an Earth orbit, the $750 million joint mission between NASA and the European Space Agency reached 34,130 mph—the greatest velocity yet required by an interplanetary probe for the task.

The scientific mission also included a controlled fire to see how flames behave in weightlessness and release of the craft's 50-foot robotic arm, which had a special material attached to estimate effects of deterioration on the Intelsat satellite, launched earlier this year into a low and useless orbit. Discovery landed at Edwards Air Force Base, California, on October 10, ending a four-day, practically flawless mission. (*W Post,* Oct 7/90; Oct 11/90; *NY Times,* Oct 7/90; Oct 11/90)

October 11: The Office of Technology Assessment warned that Space debris could make space activity too risky in 20 to 30 years if the most traveled orbits continue to be littered. With the average collision velocity in low-Earth orbit 6.2 miles per second, an object having 1/35th the weight of an aspirin tablet has the impact of a .30-caliber bullet, they said. The Air Force tracks such debris to advise NASA before Shuttle missions, but no course change has ever been deemed necessary. (*W Post,* Oct 12/90; *C Trib,* Oct 14/90)

• NASA scientists warned that examination of data from the ozone hole over the South Pole showed record depletion. They added that ozone levels throughout the Southern Hemisphere were as low as they had ever been. The ozone hole has been monitored since 1979 with the Total Ozone Mapping Spectrometer, an instrument on the Goddard-Managed NIMBUS-7 spacecraft. (*C Trib,* Oct 12/90; NASA Release 90-137)

October 16: The Senate, in a rush to agree on the 1991 spending plan, compromised on a $1.9 billion allocation for NASA's Space Station. Said Senator Barbara Mikulski, "The space program has been increased by 13 percent so it's not as if we gouged them." (W Post, Oct 17/90; *AvWk,* Oct 22/90)

October 17: NASA released a photograph from the Cosmic Background Explorer (COBE) of the vast dust clouds that fill the Milky Way galaxy. Launched almost one year previously, the COBE was to study background radiation from the Big Bang believed to have created the universe. It found that the fireball was unexpectedly smooth, belying numerous theories. (*W Post,* Oct 18/90; *NY Times,* Oct 23/90; NASA Release 90-140)

October 25: NASA announced the selection of 280 research proposals for immediate negotiation of Phase I contracts in their 1990 Small Business Innovation Research program (SBIR). SBIR had aimed at stimulating technological innovation in small business for the past eight years; the Phase I contracts normally would not exceed $50,000. (NASA Release 90-144)

October 26: NASA considered a proposal for a major repair of the Hubble Space Telescope that would include the reinstating of full observation abilities. In addition to replacing the telescope's main camera in 1993, as already planned, astronauts would also replace another device with corrective optics for light that reached other important scientific instruments. The plan called for repairing or replacing the craft's solar panels as well, because their vibrations had interfered with many sensitive observations. (*NY Times*, Oct 27/90)

October 30: NASA declared its Shuttle fleet "leak free" as Columbia passed a fueling test after two defective seals had been replaced and nuts and connectors in the engine compartment had been tightened. Space Shuttle Atlantis had passed its fuel test a week earlier and was scheduled for liftoff November 9. (*NY Times*, Oct 31/90; *W Post*, Oct 31/90; NASA Release N90-86)

• A Delta rocket was launched from Cape Canaveral, Florida, successfully carrying an International Maritime Satellite Organization communications satellite into orbit. The satellite was intended to expand international telephone and facsimile services aboard ships, planes, trains, and other vehicles. (W Post, Oct 31/90)

During October: Photographs sent back from the Magellan spacecraft since September 15 were 10 times sharper than any seen before. The images "revealed a whole new planet" and "almost everywhere you look on Venus you see volcanic features," said scientists at the Jet Propulsion Laboratory in California. (*CSM*, Oct 25/90)

• United States and Soviet space and aeronautics officials planned three separate meetings for October to visit U.S.S.R. space facilities. A U.S./U.S.S.R. Solar System Exploration Joint Working Group meeting was also planned for October in Virginia, where the group would visit NASA's aeronautics centers. (NASA Release 90-138)

• NASA selected ST Systems Corporation for negotiation to a five-year, $37 million contract for operation and analysis support for Goddard's National Space Science Data Center, Greenbelt, Maryland. NASA also awarded CBI Services, Inc. a $28.2 million contract for work on Ames Research Center's 12-foot pressure wind tunnel. (NASA Release C90-dd; C 90-ee)

• NASA announced in October that its expertise in structural mechanics, design optimization, and computer codes was finding application in orthopedic science in the custom design of prosthetic joint implants. Similar development of a computer code, said experts, would allow implants a life span increase from an average of 7 years to 20 years while providing less discomfort to the patient. (NASA Release 90-142)

November

November 4: NASA officials announced that the Hubble Space Telescope for the first time had observed a faint object, a quasar billions of light years from Earth. Officials hoped that study of the image, known to astronomers as UM675, would give clues as to the chemicals present soon after the universe was born. (*W Times,* Nov 5/90)

November 5: Arnold Aldrich, Associate Administrator for Aeronautics, Exploration and Technology, announced an organizational change at the Ames-Dryden Flight Research Center, Mountain View, California, designed to reinforce and strengthen national flight research capabilities at the Ames-Dryden Flight Research Facility (DFRF), Edwards, California. Kenneth J. Szalai was chosen as the Ames Deputy Director, effective December 3, 1990. He was, however, located at Dryden and directed research and operations. Szalai previously served as Chief of the Research Engineering Division at Dryden. (NASA Release 90-149)

November 7: NASA officials said that they had begun overhauling the design of their Space Station, a response to budget cuts passed by Congress and a directive to eliminate $6 billion from the Space Station plan over the next five years. William Lenoir, NASA's associate administrator for space flight, said that the redesign would make the Space Station an orbiting laboratory merely visited by astronauts for scientific research. He added that it would not only be stripped of equipment for studying Earth and to make astronomical observations, but for servicing spacecraft on their way to the Moon and Mars. (*WSJ,* Nov 8/90; Nov 9/90; Nov 12/90; *W Post,* Nov 8/90; *NY Times,* Non 9/90)

November 12: A Titan 4 rocket blasted off from Cape Canaveral, Florida, with a classified payload outside experts believed was a $180 million advanced missile warning satellite. It was the third launch of the Titan 4 and followed a two-month delay caused by undisclosed technical problems. The satellite, they added, would likely observe Iraq. The Space Shuttle Atlantis, which occupied a nearby launch pad for liftoff on the November 15, also had a classified cargo aboard thought to be for the same purpose. (*W Post,* Nov 13/90; *W Times,* Nov 13/90)

November 13: The Galileo spacecraft, launched from Space Shuttle Atlantis October 18, 1989, fired its thrusters so it could come closer to Earth and use our planet's gravity to slingshot in December toward the sun for a 1995 rendezvous with Jupiter. The fly-by would mark the first time Earth had been approached by a spacecraft from Venus, or anywhere inside Earth's orbit. The craft was scheduled to study Earth during the maneuver, slingshot the sun and study our planet again on the return trip in 1992, and then study Jupiter's atmosphere for two more years after reaching the planet. The loop around the solar system was necessary because the rocket that launched the craft was too weak to propel Galileo directly on its two-year trip to Jupiter. (*Fla Today,* Nov 14/90; *P Inq,* Nov 14/90)

• NASA selected investigators and science teams from 11 U.S. universities, 3 NASA centers, 3 other U.S. laboratories, and 13 foreign countries for the Saturn orbiter portion of the Cassini mission scheduled for launch in 1996. The mission would include 62 investigations for analysis of Saturn's atmosphere, its ring particles, and moons within them. The orbiter would also send a probe to Saturn's moon, Titan, and later map it. (NASA Release 90-150)

November 15: Space Shuttle Atlantis, flight STS-38, was launched from Cape Canaveral, Florida with a classified satellite, probably to be used for reconnaissance in the Persian Gulf region. The flight marked the last time a Shuttle would be assigned a secret and sole military payload because the Pentagon planned to use untended rockets for classified missions in the future. The satellite was deployed the following day.

Atlantis touched down at Cape Canaveral on November 20 after a continuation of high winds and rains had prevented landing at Edwards Air Force Base in California the day before. NASA preferred using the seven-mile runway at Edwards, as opposed to the three-mile strip at Kennedy Space Center in Florida, which had not been used since 1985. Astronauts, however, said they felt no hesitation in using the Kennedy runway. This was only the sixth time the orbiter had used the runway and the first time since the Challenger accident. Two more military missions were planned for the Space Shuttle, but the payloads would be unclassified. (*NY Times,* Nov 16/90; Nov 21/90; *W Post,* Nov 16/90; Nov 21/90; NASA Anno)

November 26: The Air Force launched a Delta 2 rocket from Cape Canaveral, Florida, that carried a $65 million NAVSTAR Global Positioning System satellite into orbit. The navigation satellite was 10th in a series of 21 that the United States hoped to have in orbit by 1992. (*W Post,* Nov 27/90; *USA Today,* Nov 27/90)

• NASA announced the selection of 84 proposals for Phase II contracts in NASA's Small Business Innovation Research Program. Aiming to stimulate technological innovation from small business, the contracts were allowed to reach $500,000 over a two-year period. (NASA Release 90-152)

November 27: The six-member Hubble Optical Systems Board of Investigation, formed by NASA and headed by Dr. Lew Allen of the Jet Propulsion Laboratory at Pasadena, California, released its final report, cumulating a five-month study. The report stated that both NASA and the Perkin-Elmer Corporation, then known as Hughes Danbury Optical Systems, were responsible for the flawed mirror that prevented the Hubble Space Telescope from distinguishing very faint objects. Lower management at the Corporation discounted any information that conflicted with certified procedures for testing the mirror with newly designed equipment. Upper management at Perkin-Elmer and at NASA, the report went on, were left unaware of potential problems the mirror might have had in 1981. Dr. Allen and Leonard Fisk, NASA's Chief Scientist, agreed that NASA management operated under different procedures then and would have caught the problem in 1990. (*NY Times,* Nov 28/90; *W Post,* Nov 28/90)

During November: NASA and the Technology Utilization Foundation held their first "Technology 2000" conference in Washington, D.C., to encourage marketing of space technology. Among the innovations showcased were an implant device for releasing insulin in programmed doses; a light sensing instrument called the "Video Harp"; a system to allow the hearing impaired to locate loud noises such as sirens; a digital audio signal processing system for fuller, more realistic sound called the "Convolvotron"; the model 700 Infrared Temperature Taking System; and the Intelligent Physics Tutor with step-by-step instructions to a student. High-tech gear only a few years away, said USA Today, included a space-age washer and dryer, refrigerator, and gloves. (*USA Today,* Nov 29/90)

• NASA's Langley Research Center at Langley Field, Hampton, Virginia, and Honeywell's Space System Group, successfully concluded a joint flight research project that involved study of Global Positioning System (GPS) satellites for improved automated landing in spacecraft and aircraft. Fifteen GPS satellites were in orbit at the time of the test, and the group hoped to add nine more by 1993.

Meanwhile, NASA and the Technical and Administrative Services Corporation signed a two-year agreement for the exchange of information associated with closed environment systems related to food production both on Earth and in space. The hydroponic technology sharing involved no exchange of funds. (NASA Release 90-155, 90-148)

• NASA convened with the Space Activities Commission of Japan in November for the fourth annual meeting of the NASA/SAC Cooperative Space Activities Planning Group in Arlington, Virginia. Discussions focused on a number of collaborative projects in the fields of astrophysics, solar system science, microgravity science, life science, and Earth observation. (NASA Release 90-156)

• Physicians at the Johnson Space Center, Houston, Texas, announced in November that they had instituted a new treatment, promethazine, for space motion sickness. Since NASA's return to flight in September 1988, the medication had helped decrease the malady. (NASA Release 90-155)

December

December 2: Space Shuttle Columbia (flight STS-35) was launched from Cape Canaveral, Florida, carrying the $150 million Astro-1 for a 10-day astronomy mission. The payload consisted of one x-ray telescope and three ultraviolet telescopes that would supplement and spot targets for the Hubble Space Telescope. The project originally had been scheduled for a May 30 liftoff, but subsequent hydrogen leaks grounded the entire fleet. The Astro-1 crew had the longest wait for this mission in Space Shuttle history. It was to have observed Halley's Comet in 1986, but the mission was postponed after the Challenger accident. The x-ray telescope was controlled from Goddard and the three ultraviolet instruments from Marshall Space Flight Center, Huntsville, Alabama.

Columbia landed at Edwards Air Force Base in California on December 11, one day earlier than planned to avoid a rain forecast. The Astro-1 mission did not achieve all of its objectives because of a failed computer that overheated from lint in the cooling system. Because astronomers on board were forced to aim the telescopes manually at predetermined objects, only 135 of the slated 250 targets were observed. (*NY Times,* Dec 3/90; Dec 7/90; Dec 12/90; *W Post,* Dec 3/90; Dec 7/90; Dec 12/90)

• A Soviet rocket carrying a TM-11 space capsule was launched into orbit from Kazakhstan Republic in Central Asia on this date. Accompanying the two cosmonauts on board was a Japanese television reporter, making him the first journalist in space. Japan's biggest private television company, TBS, as well as other Japanese companies, had their slogans emblazoned on the rocket (*B Sun,* Dec 3/90; *LA Times,* Dec 3/90)

December 4: The Ulysses spacecraft launched from Space Shuttle Discovery and developed a problem on its trip to rendezvous with Jupiter in 1992, from which it would slingshot to achieve polar orbit around the sun. It was believed that the difficulty came from a 24-1/2-foot antenna boom meant to study radio signals in space and waves in the solar winds. Scientists ascertained that the boom was bent, perhaps from the sun's rays, and was causing the craft to wobble. Because of the wobble, the main dish-shaped antenna would not be able to focus on Earth and relay data gleaned from the star. Scientists were confident that the problem would be solved before the craft's 1994 mission began. (*C Trib,* Dec 15/90; *NY Times,* Dec 16/90)

December 8: The Galileo spacecraft came to within 590 miles of Earth and received a push from Earth's gravity that increased its speed from 67,000 m.p.h. to 78,000 m.p.h. This was the first of two gravity assisted passes by Earth that were planned for the craft, enabling it to reach Jupiter in December 1995. It collected numerous data from Earth, as it would again on its second pass. (*CSM,* Dec 10/90; *NY Times,* Dec 9/90)

December 10: A White House commission, formed in July as a result of the Hubble Space Telescope problem and temporary grounding of the Shuttle fleet, released its report concerning future space plans. The 12-member panel emphasized a new system of crewless launch vehicles for a number of NASA missions in order to reduce reliance on the Space Shuttle. It also recommended a greater pursuance of scientific study and less emphasis on the accomplishment of engineering feats for their own sake, that is, crew-assisted space flight. In addition, it recommended a gradual phasing out of the Space Shuttle; redesign of the Space Station to make it cheaper and simpler, with a focus on life in space; and restructuring of Agency management.

William Lenoir, head of NASA's space flight program, said that a heavy-lift rocket would take $10 billion to design and take 7 to 10 years to complete. He also said that new Space Station designs were underway, but would be similar to the original ones. (*WSJ,* Dec 7/90; Dec 11/90; *NY Times,* Dec 11/90; *W Post,* Dec 11/90; *H Post,* Dec 14/90; NASA Release 90-161)

December 20: The selection of 12 research projects, under the auspices of the Earth Observations Commercialization Applications Program, for developing private sector applications of space-based and airborne remote sensing technologies, were selected by NASA. The 12 projects would lead to 1-year funding contracts and would identify and research new commercial products and services that might be developed from existing technology. (NASA Release 90-164)

December 26: NASA announced its Integrated System Preliminary Design Review for Space Station Freedom was completed. Preliminary figures for June 1990 showed Freedom's weight to be 143,000 pounds lighter than the allocated limit of 512,000 pounds and that the housekeeping power limit of 45 kilowatts was exceeded by 15 kilowatts. A summer-long resources scrub reduced weight estimates by 130,000 pounds and housekeeping power by 13 kilowatts. "What this means," said Deputy Manager Marc Bensimon, "is that the design is maturing and converging." (NASA Release 90-165)

During December: NASA awarded contracts to Centennial Contractors, SPACEHAB, and Dyncorp. Centennial was given a one-year, $5 million contract for various minor constructions, modifications, and rehabilitation projects at the Goddard Space Flight Center, Greenbelt, Maryland, SPACEHAB received a five-year, $184,236,000 contract for providing commercial mid-deck augmentation module services: physical and operational integration of the module and the experiments; power, cooling, data management; and crew training in six Space Shuttle flights at six-month intervals. Dyncorp procured a $30 million contract for technical services that included operation and maintenance of radar, telecommunications, and optical services. (NASA Release C90-gg; 90-157; *W Times*, Dec 4/90)

• Colonel Robert C. Springer retired from NASA and the Marine Corps. He was both an astronaut and a mission specialist and planned to work for Boeing Aerospace. (NASA Release 90-159)

During 1990: The year saw both major scientific achievements and several disappointments. The Hubble Space Telescope was successfully deployed from Space Shuttle Discovery in April but was discovered to have a spherical aberration that prevented the most distant observations for which it was designed. However, it began unprecedented scientific work in spectroscopy, photometry, astronomy, and ultraviolet wavelength imaging not possible from the ground. The telescope also sent back impressive photographs from Orion's nebula and the giant Saturn storm by using computer image processing.

The Magellan spacecraft began detailed mapping of Venus in August, the European Space Agency's Ulysses spacecraft was launched in October by Space Shuttle Discovery to study the poles of the Sun, and the Galileo spacecraft made its first gravity-assisted pass of the Earth in December. Meanwhile, NASA Administrator Richard Truly launched an effort to collect the best ideas on how to return to the Moon and go to Mars.

Six successful Shuttle missions were flown, a standdown of five months because of hydrogen leaks notwithstanding. Two Department of Defense payloads, a SYNCOM IV communications satellite, the Hubble Space Telescope, and the Ulysses spacecraft were deployed, and the Long Duration Exposure Facility satellite was retrieved. The Astro-1 astronomy mission was also successfully completed, though not to every scientist's satisfaction. (NASA Release, 90-160)

Appendix A

SATELLITES, SPACE PROBES, AND HUMAN SPACE FLIGHTS, 1986–1990

Launch Date (GMT), Spacecraft Name, COSPAR Designation, Launch Vehicle	Mission Objectives, Spacecraft Data	Apogee and Perigee (km), Period (min.) Inclination to Equator (º)	Remarks
Jan. 12 Space Shuttle Columbia (STS-61C) 3A	Objective: To launch RCA Satcom K-1 and to conduct experiments. Spacecraft: Shuttle orbiter carrying satellite as well as experiments: Materials Science Laboratory-2 (MLS-2), Hitchhiker G-1, Particle Analysis Cameras for the Shuttle (PACS), Capillary Pump Loop (CPL), Shuttle Environment Effects on Coated Mirrors (SEECM), 12 experiments flown on Get Away Special (GAS) Bridge Assembly, additional GAS experiment, and Infrared Imaging Experiment (IR-IE). Middeck payloads: Initial Blood Storage Experiment (IBSE), Comet Halley Active Monitoring Program (CHAMP), and 3 Space Science Student Involvement Program (SSIP) experiments. Weight: 12,708 lbs.	350.0 327.0 91.3 28.5	Twenty-fourth flight of Space Transportation System (STS).Piloted by Robert L. "Hoot" Gibson and Charles F. Bolden, Jr. Mission specialists Franklin R. Chang-Diaz, Steven A. Hawley, and George D. "Pinky" Nelson. Payload specialists Robert J. Cenker (RCA) and Congressman C. William "Bill" Nelson, first member of U.S. House of Representatives in space. Columbia launched from KSC at 6:55 a.m. EST on Jan. 18. Mission duration: 6 days, 2 hrs, 4 min.
Jan. 12 RCA Satcom K-1 3B	Objective: To launch communications satellite. Spacecraft: Box-shaped, 67x84x60 inch main structure, three-axis stabilized, twin 280 square ft. solar panels deployed after launch. Weight at launch: 15,929 lbs.	35,794.0 35,781.0 1,436.2 0.0	Second in series of satellites for RCA American Communications, Inc.Deployed from Columbia.
Jan. 28 Space Shuttle Challenger (STS-51L)	Objective: To deploy/launch TDRS-B/IUS (Tracking and Data Relay Satellite/Inertial Upper Stage), deploy/retrieve the SPARTAN-203/Halley (Shuttle Pointed Autonomous Research Tool for Astronomy), and to conduct experiments. Spacecraft: Shuttle orbiter carrying satellites and experiments. Payloads include: Hughes Aircraft Company Fluid Dynamics Experiment (FDE), composed of six experiments; second flight of the Comet Halley Active Monitoring Program (CHAMP); the Phase Partitioning Experiment (FPE) to dissolve two polymer solutions in water to	N/A	Challenger failed 73 seconds into the mission

Launch Date (GMT), Spacecraft Name, COSPAR Designation, Launch Vehicle	Mission Objectives, Spacecraft Data	Apogee and Perigee (km), Period (min.) Inclination to Equator (º)	Remarks
	observe separation; six experiments for the Teacher in Space, including hydrophonics, magnetism, Newton's laws, effervescence, chromatography, and simple machines; and three Shuttle Student Involvement Program (SSIP) experiments.		
Feb. 9 Defense 14A Atlas H	Objective: Development of space flight techniques and technology. Spacecraft: Not announced.	Not announced	Achieved orbit.
Feb. 9 Defense 14E	Objective: Development of space flight techniques and technology. Spacecraft: Not announced.	Not announced	Achieved orbit.
Feb. 9 Defense 14F	Objective: Development of space flight techniques and technology. Spacecraft: Not announced.	Not announced	
Feb. 9 Defense 14H	Objective: Development of space flight techniques and technology. Spacecraft: Not announced.	Not announced	Achieved orbit.
Mar. 13 Soyuz T-15 USSR 1986-022A			Crew consisted of Leonid Kizim and Vladimir Solovyov. Docked with MIR space station on May 5-6 and transferred to Salyut 7 complex. Transferred from Salyut 7 back to MIR on Jun. 25-26. Flight time: 125 days, 1 hr, 1 min.
Sept. 5 Defense 69A Delta 180	Objective: Development of space flight techniques and technology. Spacecraft: Not announced.	719 212 93.9 39.1	Launched by NASA for DoD's Strategic Defense Initiative (SDI) space intercept and collision test. Destroyed.
Sept. 5 Defense 69B	Objective: Development of space flight techniques and technology. Spacecraft: Not announced.	611 223 92.9 22.8	Strategic Defense Initiative (SDI) space intercept and collision test. Destroyed.

Launch Date (GMT), Spacecraft Name, COSPAR Designation, Launch Vehicle	Mission Objectives, Spacecraft Data	Apogee and Perigee (km), Period (min.) Inclination to Equator (º)	Remarks
Sept. 17 NOAA 10 73A Atlas E	Objective: To launch spacecraft into sun-synchronous orbit to accomplish operational mission requirements. Spacecraft: Launch configuration: 4.9m high, 1.9m in diameter. Weight: 1,712 kg.	823 804 101.2 98.7	Third in advanced Tiros-N series. Funded by National Oceanic and Atmospheric Administration (NOAA) and launched by NASA. Also onboard Search and Rescue instruments.
Nov. 14 Polar Bear 88A Scout	Objective: To place Air Force P87-1 satellite into orbit to achieve mission objectives. Spacecraft: Not announced. Weight: 270 lbs.	1,015 960 104.9 89.6	Launched by NASA for Air Force. Reconditioned satellite that hung for several years in National Air and Space Museum. Experiments to study radio interference caused by Aurora Borealis, or Northern Lights.
Dec. 5 Fltsatcom 7 96A Atlas-Centaur	Objective: To launch satellite into planned geostationary position. Spacecraft: Hexagonal, composed of payload and spacecraft module; 22.7 ft high. Provides 1 EHF and 23 UHF communications channels. Weight at liftoff: 5,073 lbs. Weight in orbit: 2,488 lbs.	36,024 35,551 1,436.2 5.2	Launched by NASA for Navy; to serve DoD.

Launch Date (GMT), Spacecraft Name, COSPAR Designation, Launch Vehicle	Mission Objectives, Spacecraft Data	Apogee and Perigee (km), Period (min.) Inclination to Equator (°)	Remarks
Feb. 5 Soyuz TM-2 USSR 1987-013A			Crew consisted of Yuriy Romanenko and Aleksandr Laveykin. Docked with MIR space station. Romanenko established long-distance stay in space record of 326 days. Flight time: 174 days, 3 hrs, 26 min.
Feb. 12 Defense 15A Titan 3B	Objective: Development of space-flight techniques and technology. Spacecraft: Not announced.	Not announced	Achieved orbit
Feb. 26 GOES 7 22A Delta 179	Objective: To launch satellite into planned geostationary position. Spacecraft: Cylindrical, 2.2m in diameter and 4.5m long from top of S-band omnidirectional antenna mast to bottom of apogee boost motor. Apogee boost motor eject-ed after synchronous orbit achieved. Weight at liftoff: 840 kg. Weight in geostationary position: 456 kg.	35,796 35,783 1,436.3 0.1	Eighth in series of operational environ-ment monitoring satel-lites. Launched by NASA for National Oceanic and Atmos-pheric Administration (NOAA). Turned over to NOAA Mar. 25 after successful apogee motor fire Feb. 27. Spacecraft joins GOES 6 (WEST) as GOES 7 (EAST) part of a two-satellite operational system.
Mar. 20 Palapa B-2P 29A Delta 182	Objective: To launch communica-tions satellite. Spacecraft: Drum-shaped, telescoping cylinder, with antennas deployed 22.1 ft high, and 7 ft in diameter. Weight at liftoff: 7,515 lbs. Weight in geo-synchronous orbit: 1,437 lbs.	35,788 35,788 1,436.2 0.0	Fifth in series of Indonesian communi-cations satellites. Launched by NASA for Republic of Indonesia. Will replace Palapa B2 launched in Feb. 1984 by orbiter Challenger, which malfunctioned. Palapa B-2P to join Palapa B1 orbited by Shuttle in Jun. 1983. Part of two-satellite system providing ser-vice to more than 13,000 islands.

Launch Date (GMT), Spacecraft Name, COSPAR Designation, Launch Vehicle	Mission Objectives, Spacecraft Data	Apogee and Perigee (km), Period (min.) Inclination to Equator (º)	Remarks
May 15 Defense 43A Atlas H	Objective: Development of space flight techniques and technology. Spacecraft: Not announced.	Not announced	Achieved orbit.
Jun. 20 DMSP F-8 53A Atlas E	Objective: To launch meteorological observation satellite into planned orbit. Spacecraft: Same basic configuration as DMSP F-6. Weight: 1,421 kg.	857 835 101.9 98.8	Third satellite in Block 5D-2 series. Data also used by National Oceanic and Atmospheric Administration (NOAA).
Jul. 22 Soyuz TM-3 USSR 1987-063A			Crew consisted of AleksandrViktorenko, Aleksandr Aleksandrov, and Mohammed Faris. Docked with Mir space station. Aleksandrov (first Bulgarian in space) remained in MIR 160 days; returned with Yuriy Romanenko. Viktorenko and Faris returned Jul. 30 in Soyuz TM-2, with Aleksandr Laveykin, who experienced medical problems. Faris first Syrian in space. Flight time: 160 days, 7 hrs, 16 min.
Sep. 16 Defense 80A Scout	Objective: To place satellite into orbit to achieve Navy objectives. Spacecraft: Not announced.	1,175 1,017 107.2 90.3	First of two Transit satellites launched by Scout vehicle. Launched by NASA for DoD.
Sep. 16 Defense 80B	Objective: To place satellite into orbit to achieve Navy objectives. Spacecraft: Not announced.	1,181 1,014 107.2 90.3	Second of two Transit satellites launched by NASA for DoD on same Scout.
Oct. 26 Defense 90A Titan 34D	Objective: Development of space flight techniques and technology. Spacecraft: Not announced.	Not announced	Achieved orbit.

Launch Date (GMT), Spacecraft Name, COSPAR Designation, Launch Vehicle	Mission Objectives, Spacecraft Data	Apogee and Perigee (km), Period (min.) Inclination to Equator (º)	Remarks
Nov. 29 Defense 97A Titan 34D	Objective: Development of space-flight techniques and technology. Spacecraft: Not announced.	Not announced	Achieved orbit.
Dec. 21 Soyuz TM-4 USSR 1987-104A			Crew consisted of Vladimir Titov, Musa Manarov, and Anatoliy Levchenko. Docked with MIR space station. Crew of Yuriy Romanenko, Aleksandr Aleksandrov, and Anatoliy Levchenko returned Dec. 29 in Soyuz TM-3. At end of 1987 still docked with MIR. Flight time: 180 days, 5 hrs

Launch Date (GMT), Spacecraft Name, COSPAR Designation, Launch Vehicle	Mission Objectives, Spacecraft Data	Apogee and Perigee (km), Period (min.) Inclination to Equator (º)	Remarks
Feb. 2 DMSP F-9 6A Atlas E	Objective: To launch meteorological observation satellite into planned orbit. Spacecraft: Same basic configuration as DMSP F-6. Weight: 1,421 kg.	826 815 101.3 98.7	Achieved orbit.
Feb. 8 8A DoD-2 Delta 181	Objective: To demonstrate effective ways to discriminate between decoys and reentry vehicles. Spacecraft: Not announced.	333 223 90.1 28.6	Launched by NASA for Strategic Defense Initiative (SDI). Successful test of technology. Reentered Mar. 1.
Mar. 25 San Marco D/L 26A Scout	Objective: To launch satellite into low-earth orbit to explore relationship between solar activity and meteorological phenomena by studying dynamic processes occurring in troposphere, stratosphere, and thermosphere. Spacecraft: Spherical, 96.5 cm in diameter with four canted 48 cm monopole antennas for telemetry and command. Weight: 237 kg.	615 263 93.4 3.0	Launched by NASA as joint research mission with Italian Space Commission. Launch took place from San Marino Range. Reentered over central Africa and was consumed Dec. 6.
Apr. 25 SOOS-3 33D Scout	Objective: To place Navy satellite into orbit to achieve Navy objectives. Spacecraft: Not announced. Weight: 141 lbs.	Not announced	Launched by NASA for Navy. Stacked Oscar on Scout (SOOS) part of Navy's long-established continuous all-weather global navigation system.
Apr. 25 S00S-3 33E Scout	Objective: To place Navy satellite into orbit to achieve Navy objectives. Spacecraft: Not announced. Weight: 141 lbs.	Not announced	Second of two satellites launched by NASA for Navy.
Jun. 7 Soyuz TM-5 USSR 1988-048A			Crew consisted of Viktor Savinykh, Anatoly Solovyev, and Aleksandr Aleksandrov. Docked with Mir space station. Crew returned Jun. 17 in Soyuz TM-4. Flight time: 9 days, 20 hrs, 13 min.

Launch Date (GMT), Spacecraft Name, COSPAR Designation, Launch Vehicle	Mission Objectives, Spacecraft Data	Apogee and Perigee (km), Period (min.) Inclination to Equator (º)	Remarks
Jun. 16 NOVA-II 52A Scout	Objective: To place Navy satellite into orbit to achieve Navy objectives. Spacecraft: Cylindrical main body, with four solar panels extending from main body; extendable boom from central part of satellite. Weight: 375.8 lbs.	1,199 1,150 108.9 90.0	Third in series of improved Transit navigation satellites launched by NASA for Navy.
Aug. 25 SOOS-4 74A Scout	Objective: To place Navy satellite into orbit to achieve Navy objectives. Spacecraft: Not announced. Weight: 141 lbs.	1,176 1,032 107.4 90.0	Launched by NASA for Navy into polar orbit. Stacked Oscar on Scout (SOOS).
Aug. 25 SOOS-4 74B Scout	Objective: To place Navy satellite into orbit to achieve Navy objectives. Spacecraft: Not announced. Weight: 141 lbs.	1,178 1,032 107.4 90.0	Second of two satellites launched by NASA for Navy.
Aug. 29 Soyuz TM-6 USSR 1988-075A			Crew consisted of Vladimir Lyakhov, Valeriy Polyakov, and Abdul Mohmad. Docked with MIR space station. Mohmad first Afghan in space. Crew returned Sep. 7 in Soyuz TM-5. Flight time: 8 days, 19 hrs, 27 min.
Sep. 2 DoD 77A Titan 34D		Not announced	Achieved orbit.
Sep. 5 DoD 78A Titan 2		Not announced	Achieved orbit.

Launch Date (GMT), Spacecraft Name, COSPAR Designation, Launch Vehicle	Mission Objectives, Spacecraft Data	Apogee and Perigee (km), Period (min.) Inclination to Equator (°)	Remarks
Sep. 24 89A NOAA-11 Atlas-E	Objective: To launch satellite into sun-synchronous orbit. Spacecraft: Launch configuration 491 cm high, 188 cm in diameter. Weight: 1,712 kg. Weight in orbit: 1,038 kg.	863845 102.0 98.9	Fourth of advanced TIROS-N spacecraft. Funded by National Oceanic and Atmospheric Administration (NOAA) and launched by NASA. To replace NOAA-9 as afternoon satellite in NOAA's two-polar-satellite system. Also onboard Search and Rescue instruments.
Sep. 29 Space Shuttle Discovery (STS-26) 91A	Objective: To deploy Tracking and Data Relay Satellite-C/Inertial Upper Stage.Spacecraft: Shuttle orbiter carrying satellite as well as experiments: Automated Directional Solidification Furnace (ADSF), Aggregation of Red Blood Cells (ARC), Earth Limb Radiance Experiment (ELRAD), Isoelectric Focusing Experiment (IFE), Infrared Communication Flight Experiment (IRCFE), Mesoscale Lightning Experiment (MLE), Protein Crystal Growth (PCG), Phased Partitioning Experiment (PPE), and Physical Vapor Transport of Organic Solids (PVTOS). Additionally, 3 Shuttle Student Involvement Project (SSIP) experiments. Weight: 38,774 lbs.	336 306 91.0 28.5	Twenty-sixth flight of Space Transportation System (STS). First since Challenger accident of Jan. 28, 1986. Piloted by Frederick H. Hauck and Richard O. Covey. Mission specialists John M. Lounge, David C. Hilmers, and George D. Nelson. Discovery launched from KSC at 11:37 a.m. EDT. Satellite deployed and experiments conducted. Shuttle landed at Edwards AFB, CA, at 12:37 p.m. EDT on Oct. 3. Mission duration: 4 days, 1 hr.

Launch Date (GMT), Spacecraft Name, COSPAR Designation, Launch Vehicle	Mission Objectives, Spacecraft Data	Apogee and Perigee (km), Period (min.) Inclination to Equator (º)	Remarks
Sep. 29 TDRS 3 91B	Objective: To deliver TDRS satellite to stationary geosynchronous orbit. Spacecraft: Three-axis stabilized, momentum-biased configuration with two sun-oriented solar panels attached. TDRS measures 57.2 ft tip-to-tip of deployed solar panels. TDRS composed of 3 modules: (1) equipment module houses attitude control, electrical power, propulsion, telemetry, tracking, and command subsystems; (2) payload module consists of processing and frequency-generation equipment; and (3) antenna module supports dual deployable and fixed antennas, multiple-access array, and remainder of telecommunications hardware. Weight at launch, including Inertial Upper State (IUS) booster: 37,699 lbs. Weight in orbit: 4,637 lbs.	35,803 35,779 1436.3 0.2	Third in series. Launched from Shuttle orbiter Discovery. *Completes two-satellite constellation, located at 171° W. Leased by NASA from Continental Telephone Company (CONTEL).*
Nov. 6 DoD 99A Titan 34D	Objective: Development of spaceflight techniques and technology. Spacecraft: Not announced.	Not announced	Achieved orbit.
Nov. 26 Soyuz TM-7 USSR 1988-104A			Crew consisted of Aleksandr Volkov, Sergey Krikalev, and Jean-Loup Chretien. Docked with MIR space station. Volkov and Krikalev in orbit at end of 1988. Soyuz TM-6 returned with Chretien, Vladimir Titov, and Musa Manarov. Titov and Manarov completed 366-day mission on Dec. 21. Flight time: 151 days, 11 hrs.

Launch Date (GMT), Spacecraft Name, COSPAR Designation, Launch Vehicle	Mission Objectives, Spacecraft Data	Apogee and Perigee (km), Period (min.) Inclination to Equator (°)	Remarks
Sep. 29 TDRS 3 91B	Objective: To deliver TDRS satellite to stationary geosynchronous orbit. Spacecraft: Three-axis stabilized, momentum-biased configuration with two sun-oriented solar panels attached. TDRS measures 57.2 ft tip-to-tip of deployed solar panels. TDRS composed of 3 modules: (1) equipment module houses attitude control, electrical power, propulsion, telemetry, tracking, and command subsystems; (2) payload module consists of processing and frequency-generation equipment; and (3) antenna module supports dual deployable and fixed antennas, multiple-access array, and remainder of telecommunications hardware. Weight at launch, including Inertial Upper State (IUS) booster: 37,699 lbs. Weight in orbit: 4,637 lbs.	35,803 35,779 1436.3 0.2	Third in series. Launched from Shuttle orbiter Discovery. Completes two-satellite constellation, located at 171° W. Leased by NASA from Continental Telephone Company (CONTEL).
Nov. 6 DoD 99A Titan 34D	Objective: Development of spaceflight techniques and technology. Spacecraft: Not announced.	Not announced	Achieved orbit.
Nov. 26 Soyuz TM-7 USSR 1988-104A			Crew consisted of Aleksandr Volkov, Sergey Krikalev, and Jean-Loup Chretien. Docked with MIR space station. Volkov and Krikalev in orbit at end of 1988. Soyuz TM-6 returned with Chretien, Vladimir Titov, and Musa Manarov. Titov and Manarov completed 366-day mission on Dec. 21. Flight time: 151 days, 11 hrs.

Launch Date (GMT), Spacecraft Name, COSPAR Designation, Launch Vehicle	Mission Objectives, Spacecraft Data	Apogee and Perigee (km), Period (min.) Inclination to Equator (º)	Remarks
Dec. 2 Space Shuttle Atlantis 106A	Objective: To launch classified DoD Payload. Spacecraft: Shuttle orbiter carrying DoD satellite.	453 443 93.6 57.0	Twenty-seventh flight of Space Transportation System (STS). Piloted by Robert "Hoot" Gibson and Guy Gardner. Mission specialists Mike Mullane, Jerry Ross, and Bill Shepherd. Atlantis launched from KSC at 9:30 a.m. EST. Satellite deployed. Shuttle landed at Edwards AFB, CA, at 3:37 p.m. PST on Dec. 6. Mission duration: 4 days, 9 hrs, 6 min.
Dec. 2 DoD 106B	Objective: Development of spaceflight techniques and technology. Spacecraft: Not announced.	Not announced	Achieved orbit.

Launch Date (GMT), Spacecraft Name, COSPAR Designation, Launch Vehicle	Mission Objectives, Spacecraft Data	Apogee and Perigee (km), Period (min.) Inclination to Equator (º)	Remarks
Feb. 14 GPS (Block IIR) 13A Delta II	Objective: To launch spacecraft into orbit. Spacecraft: Not announced.	20,346 20,018 718.0 55.1	First in series of Block II operational NAVSTAR Global Positioning Satellites (GPS) launched aboard newest Air Force expendable launch vehicle, Delta II. Operational system to be composed of 21 satellites in 6 orbital planes.
Mar. 13 Space Shuttle Discovery (STS-29) 21A	Objective: To deploy Tracking and Data Relay Satellite (TDRS)-D. Spacecraft: Shuttle orbiter carrying satellite as well as experiments: Orbiter Experiments Autonomous Supporting Instrumentation System (OASIS-I), Space Station Heat Pipe Advanced Radiator Element (SHARE), Air Force Maui Optical Systems (AMOS) Calibration Test, Chromosome and Plant Cell Division in Space Experiment (CHROMEX), IMAX camera, Protein Crystal Growth (PGC), and two Shuttle Student Involvement Project (SSIP) experiments. Weight, including spacecraft and experiments: 263,289 lbs.	337 305 91.0 28.5	Twenty-eighth flight of Space Transportation System (STS). Piloted by Michael L. Coats and John E. Blaha. Mission specialists James P. Bagian, James F. Buchli, and Robert C. Springer. Discovery launched from KSC at 9:57 a.m. EST. Satellite successfully deployed and experiments conducted. Shuttle landed at Edwards AFB, CA, at 9:36 a.m. EST on Mar. 18. Mission duration: 4 days, 23 hrs, 39 min.

Launch Date (GMT), Spacecraft Name, COSPAR Designation, Launch Vehicle	Mission Objectives, Spacecraft Data	Apogee and Perigee (km), Period (min.) Inclination to Equator (°)	Remarks
Mar. 13 TDRS 4 21B	Objective: To deliver Tracking and Data Relay System (TDRS) satellite into geosynchronous orbit. Spacecraft: Three-axis stabilized, momentum-biased configuration with two sun-oriented solar panels attached. TDRS measures 57.2 ft tip-to-tip of deployed solar panels. TDRS composed of 3 modules: (1) equipment module houses attitude control, electrical power, propulsion, telemetry, tracking, and command subsystems; (2) payload module consists of processing and frequency-generation equipment; and (3) antenna module supports dual deployable and fixed antennas, multiple-access array, and remainder of telecommunications hardware. Weight at launch, including Inertial Upper Stage (IUS) booster: 43,212 lbs	499 487 94.5 47.7	Fourth in series. Launched from Shuttle Discovery. Spacecraft replaced partially degraded TDRS 1 at 41° W, and TDRS 1 moved to 79° W, serving temporarily as in-orbit spare. Leased by NASA from Continental Telephone Company (CONTEL).
Mar. 24 USA 36 26A Delta	Objective: DoD space test. Spacecraft: Not announced.	499 487 94.5 47.7	USA 36 (also known as Delta 183), experimental missile-hunting satellite equipped with laser radar, 7 video imaging cameras, and infrared imager.
May 4 Space Shuttle Atlantis (STS-30) 33A	Objective: To deploy Magellan spacecraft. Spacecraft: Shuttle orbiter carrying satellite as well as experiments: Fluids Experiment Apparatus (FEA), Mesoscale Lightning Experiment (MLE), and Air Force Maui Optical Site (AMOS) Calibration Test. Weight, including spacecraft and experiments: 260,878 lbs.		Twenty-ninth flight of Space Transportation System (STS). Piloted by David M. Walker and Ronald J. Grabe. Mission specialists Norman E. Thagard, Mary L. Cleave, and Mark C. Lee. Atlantis launched at 2:47 p.m. EDT after several countdown holds. Satellite deployed and experiments conducted. Shuttle landed at Edwards AFB, CA, at 3:43 p.m. EDT on May 8. Mission duration: 4 days, 57 minutes.

Launch Date (GMT), Spacecraft Name, COSPAR Designation, Launch Vehicle	Mission Objectives, Spacecraft Data	Apogee and Perigee (km), Period (min.) Inclination to Equator (º)	Remarks
May 4 Space Shuttle Atlantis (STS-30) 33A	Objective: To deploy Magellan spacecraft. Spacecraft: Shuttle orbiter carrying satellite as well as experiments: Fluids Experiment Apparatus (FEA), Mesoscale Lightning Experiment (MLE), and Air Force Maui Optical Site (AMOS) Calibration Test. Weight, including spacecraft and experiments: 260,878 lbs.	331 297 90.8 28.9	Twenty-ninth flight of Space Transportation System (STS). Piloted by David M. Walker and Ronald J. Grabe. Mission specialists Norman E. Thagard, Mary L. Cleave, and Mark C. Lee. Atlantis launched at 2:47 p.m. EDT after several countdown holds. Satellite deployed and experiments conducted. Shuttle landed at Edwards AFB, CA, at 3:43 p.m. EDT on May 8. Mission duration: 4 days, 57 minutes.
May 4 Magellan 33B	Objective: To place satellite carrying radar sensor into orbit around Venus to obtain data on planetary surface. Spacecraft: Three-axis stabilized. Length 6.4m; solar panels with span of 9.2 m. Composed of satellite structure, thermal control, power, attitude control, propulsion, command data and storage, and telecommunications subsystems. Weight, incuding spacecraft and Inertial Upper State (IUS) booster: 45,748 lbs.	Trans-Venus trajectory.	Launched from Shuttle Atlantis. Spacecraft placed on planned trajectory for rendezvous and orbit with Venus on Aug. 10, 1990.
May 10 USA 37 35A	Objective: Not announced. Spacecraft: Not announced.	Not announced	Achieved orbit.
Jun. 10 GPS (Block IIR) 44A Delta II	Objective: To launch spacecraft into orbit. Spacecraft: Not announced.	20,401 19,962 718.0 54.7	Second in series of Block II operational NAVSTAR Global Positioning Satellites (GPS) launched by Delta II. Completed system will have 21 operational and three spare satellites. Also known as USA 38.

Launch Date (GMT), Spacecraft Name, COSPAR Designation, Launch Vehicle	Mission Objectives, Spacecraft Data	Apogee and Perigee (km), Period (min.) Inclination to Equator (º)	Remarks
Jun. 14 USA 39 46A Titan	Objective: To launch upgraded Defense Support Program satellite (DSP-1). Spacecraft: Not announced.	Not announced	First launch of Titan 4 expendable launch vehicle.
Aug. 8 Space Shuttle Columbia (STS-28) 61A	Objective: DoD mission payload/cargo not announced. Spacecraft: Shuttle orbiter carrying classified DoD payload/cargo. Six Detailed Supplementary Objectives (DSOs) completed, including those associated with Extended Duration Orbiter (EDO) Medical Program. Weight: Not announced.	317 314 90.5 56.9	Thirtieth flight of Space Transportation System (STS). Piloted by Brewster Shaw and Richard Richards. Mission specialists James Adamson, David Leestma, and Mark Brown. Columbia launched from KSC at 8:37 a.m. EDT. Shuttle landed at Edwards AFB, CA, at 9:38 EDT on Aug. 13. Mission duration: 5 days, 1 hour.
Aug. 8 USA 40 61B	Objective: Not announced. Spacecraft: Not announced.	Not announced	Achieved orbit.
Aug. 8 USA 41 61C	Objective: Not announced. Spacecraft: Not announced.	Not announced	Achieved orbit.
Aug. 18 GPS (Block IIR) 64A Delta II	Objective: To launch satellite into orbit. Spacecraft: Not announced.	20,252 20,111 718.0 55.0	Third in series of 21 Block II operational NAVSTAR Global Positioning Satellites (GPS) launched aboard Delta II expendable launch vehicle. Also known as USA 42.
Sep. 4 USA 43 69A Titan 3	Objective: Not announced. Spacecraft: Not announced.	Not announced	Achieved orbit.

Launch Date (GMT), Spacecraft Name, COSPAR Designation, Launch Vehicle	Mission Objectives, Spacecraft Data	Apogee and Perigee (km), Period (min.) Inclination to Equator (º)	Remarks
Sep. 4 USA 43 69A Titan 3	Objective: Not announced. Spacecraft: Not announced.	Not announced	Achieved orbit.
Sep. 4 USA 44 69B	Objective: Not announced. Spacecraft: Not announced.	Not announced	Achieved orbit.
Sep. 5 Soyuz TM-8 USSR 1989-071A			Crew consisted of Aleksandr Viktorenko and Aleksandr Serebrov. Docked with MIR space station. Viktorenko and Serebrov returned Feb. 9, 1990, in Soyuz TM-8. Flight time: 166 days, 6 hrs.
Sep. 6 USA 45 72A Titan 2	Objective: Not announced. Spacecraft: Not announced.	Not announced	Achieved orbit.
Sep. 25 FltSatCom F-8 77A Atlas Centaur 68	Objective: To launch Navy communications satellite into geosynchronous transfer orbit. Spacecraft: Body 22.8 ft high and 8 ft in diameter. Parabolic antenna 16 ft in diameter with 80-inch solid center surrounded by wire mesh screen, 13.5-ft helical receiving antenna, 13 inches in diameter at base, mounted outside edge of transit antenna dish. Once in orbit, folded screen is deployed. Weight, going into orbit, with apogee kick motor: 5,100 lbs.	35,799 35,776 1,436.2 4.8	Sixth, and last, in series of geosynchronous satellites. Launched by NASA/industry team for Navy. Last in NASA inventory of Atlas Centaur rockets. Part of worldwide Navy, Air Force, and DoD communications system. Also known as USA 46.

Launch Date (GMT), Spacecraft Name, COSPAR Designation, Launch Vehicle	Mission Objectives, Spacecraft Data	Apogee and Perigee (km), Period (min.) Inclination to Equator (º)	Remarks
Oct. 18 Space Shuttle Atlantis (STS-34) 84A	Objective: To launch Galileo spacecraft toward rendezvous with Jupiter. Spacecraft: Shuttle orbiter carrying spacecraft with Inertial Upper Stage (IUS) booster with additional experiments: Shuttle Solar Backscatter Ultra-Violet (SSBUV) Instrument, Growth Hormone Concentration and Distribution (GHCD) in Plants, Polymer Morphology (PM) Experiment, Mesoscale Lightning Experiment (MLE), Sensor Technology (STEX) Experiment, Air Force Maui Optical Site (AMOS) Calibration Test, IMAX camera, and 1 Shuttle Student Involvement Project (SSIP). Weight, including space probe and experiments (not including Shuttle): 46,009 lbs.	160 160 90 34.3	Thirty-first flight of Space Transportation System (STS). Piloted by Donald E. Williams and Michael J. McCulley. Mission specialists Shannon W. Lucid, Franklin R. Chang-Diaz, and Ellen S. Baker. Atlantis launched from KSC at 12:54 p.m. EDT. Shuttle landed at Edwards AFB, CA, at 12:33 p.m. EDT on Oct. 23. Mission duration : 4 days, 23 hrs, 39 min.
Oct. 18 Galileo 84B	Objective: To launch spacecraft into trajectory toward Jupiter to allow close-range studies over period of almost two years. Spacecraft: Composed of three segments: (1) probe, weighing 355 kg (including 28 kg of scientific instruments), includes heat shield, 2.5-m parachute, six instruments together with control and data system, radio-relay transmitter, and other supporting equipment; (2) spinning main section includes propulsionmodule, communications antennas, main computers, and most support systems; and (3) final segment carries field and particle instruments. Weight of orbiter (second and third segments): 2,668 kg (including 103 kg of science instruments and 935 kg of propellant).	Trans-Jupiter trajectory.	Launched from Shuttle Atlantis. Planned arrival at Jupiterin December 1995.

Launch Date (GMT), Spacecraft Name, COSPAR Designation, Launch Vehicle	Mission Objectives, Spacecraft Data	Apogee and Perigee (km), Period (min.) Inclination to Equator (º)	Remarks
Oct. 21 GPS (Block IIR) 85A Delta II	Objective: To launch satellite into orbit. Spacecraft: Not announced.	20,283 20,079 717.9 54.7	Fourth in series of Block II operational NAVSTAR Global Positioning Satellites (GPS). Also known as USA 47.
Nov. 18 COBE 89A Delta 5920	Objective: To launch satellite to measure diffuse infrared radiation (cosmic background). Spacecraft: Rotating three-axis stabilized, length 7.2 m (stowed) and 27.5 m with solar panels deployed in orbit. Carries Differential Microwave Radiometer (DMR), Far Infrared Absolute Spectrophotometer (FIRAS), and Diffuse Infrared Background Experiment (DIRBE). Weight: 2,206 kg.	895 886 102.8 99.0	Cosmic Background Explorer (COBE) successfully launched using 184th, and last, NASA-owned Delta. COBE is 65th Explorer-class mission. Launched from Vandenberg AFB, CA. Mission life: one year planned, primary mission; one year planned, extended mission.
Nov. 23 Space Shuttle Discovery (STS-33) 90A	Objective: Not announced. Spacecraft: Shuttle orbiter carrying unannounced DoD satellite.	Not announced	Thirty-second flight of Space Transportation System (STS). Piloted by Frederick D. Gregory and John E. Blaha. Mission specialists Kathryn C. Thornton, F. Story Musgrave, and Manely L. "Sonny" Carter. Discovery launched from KSC at 7:23 p.m. EST on Nov. 22. Shuttle landed at Edwards AFB, CA, at 7:31 p.m. EST on Nov. 27. Mission duration: 5 days, 7 min.
Nov. 23 DoD 90B	Objective: Not announced. Spacecraft: Not announced.		
Dec. 11 GPS (Block IIR) 97A Delta II	Objective: To launch satellite into orbit. Spacecraft: Not announced.		Fifth in series of Block II operational Global Positioning Satellites (GPS) launched by Air Force Delta II. Also known as USA 49.

Launch Date (GMT), Spacecraft Name, COSPAR Designation, Launch Vehicle	Mission Objectives, Spacecraft Data	Apogee and Perigee (km), Period (min.) Inclination to Equator (º)	Remarks
Jan. 1 Skynet 4A 1 A Titan 3	Objective: To launch satellite. Spacecraft: Not announced. Weight: 3,320 lbs.	35,782 33,685 1,382.5 3.4	Commercial launch by Martin Marietta for British Ministry of Defence.
Jan. 1 JCSAT 2 1B	Objective: To launch satellite. Spacecraft: Not announced. Weight: 14,700 lbs.	7,191 821 180.1 0.3	Second satellite in dual payload Titan 3 launch. Satellite owned by Japanese Communications Satellite Co.
Jan. 9 Space Shuttle Columbia (STS-32) 2A	Objective: To launch Syncom IV-5, retrieve Long Duration Exposure Facility (LDEF), and conduct scientific and medical experiments while in weightless environment. Spacecraft: Shuttle orbiter carrying satellite as well as middeck experiments: American Flight Echocardiograph (AFE), Characterization of Neurospora Circadian Rhythms (CNCR), Fluids Experiment Apparatus (FEA), Latitude/Longitude Locator (L3), Protein Crystal Growth (PCG), IMAX camera, Mesoscale Lightning Experiment (MLE), and Air Force Maui Optical Site (AMOS) Calibration Test. Weight of orbiter and cargo at SRB ignition: 256,670 lbs.	342 316 90.8 28.4	Thirty-third flight of Space Transportation System (STS). Piloted by Daniel C. Brandenstein and James D. Wetherbee. Mission specialists Bonnie J. Dunbar, Marsha S. Ivins, and G. David Low. Columbia launched from KSC at 7:35 a.m. EST. After deploying Syncom IV-5, Columbia crew retrieved Long Duration Exposure Facility (LDEF) (launched by STS-41C in April 1984) at 10:16 a.m. EST on Jan. 11. After detailed photo-graphic survey, satellite berthed in cargo bay 4.5 hrs later. Shuttle landed at Edwards AFB, CA, at 1:35 a.m. PST on Jan. 20. Landing delayed one day because of ground fog at landing site. Mission duration: 10 days, 21 hrs, making this longest Shuttle flight to date.

Launch Date (GMT), Spacecraft Name, COSPAR Designation, Launch Vehicle	Mission Objectives, Spacecraft Data	Apogee and Perigee (km), Period (min.) Inclination to Equator (º)	Remarks
Jan. 9 Syncom IV-5 (Leasat-5) 2B	Objective: To launch satellite into transfer orbit. Spacecraft: Cylindrical, telescoping satellite, 15 ft long, 13 ft in diameter. Weight in cargo bay: 17,000 lbs. Weight of satellite in orbit: 3,060 lbs.	36,363 34,858 1,427.1 1.4	Fourth operational satellite in Leasat system. Launched from Shuttle Columbia for Navy. Replacement for FltSatCom spacecraft.
Jan. 24 USA-50 8A	Objective: Development of space flight techniques and technology. Spacecraft: Not announced.	20,088 718.0 54.6	Sixth navigation satellite in series. Global Positioning Satellite (GPS).
Feb. 11 Soyuz TM-9 USSR 1990-014A			Crew consisted of Anatoliy Solovyov and Aleksandr Balandin. Docked with MIR space station. Crew returned Aug. 9 in Soyuz TM-9. Flight time: 178 days, 22 hrs, 19 min.
Feb. 14 USA-51 15A Delta II	Objective: Development of space flight techniques and technology. Spacecraft: Not announced.	549 532 95.3 43	Low-power Atmospheric Compensation Experiment (LACE) satellite. Part of Strategic Defense Initiative (SDI) program.
Feb. 14 USA-52 15B	Objective: Development of space flight techniques and technology. Spacecraft: Not announced. Weight: 20,275 lbs.	470 464 93.8 43.1	Relay Mirror Experiment (RME) satellite. Part of Strategic Defense Initiative (SDI) program testing.

Launch Date (GMT), Spacecraft Name, COSPAR Designation, Launch Vehicle	Mission Objectives, Spacecraft Data	Apogee and Perigee (km), Period (min.) Inclination to Equator (º)	Remarks
Feb. 28 Space Shuttle Atlantis (STS-36) 19A	Objective: To launch DoD payload. Spacecraft: Shuttle orbiter carrying unannounced satellite.	264 248 89.4 61.9	Thirty-fourth flight of Space Transportation System (STS). Piloted by John O. Creighton and John H. Casper. Mission specialists David C. Hilmers, Richard M. Mullane, and Pierre J. Thout. Atlantis launched from KSC at 2:50 a.m. EST. Shuttle landed at Edwards AFB, CA, at 10:08 a.m. PST on Mar. 4. Mission duration: 4 days, 10 hrs, 19 min.
Feb. 28 USA-53 19B	Objective: Development of spaceflight techniques and technology. Spacecraft: Not announced.	Not announced	Classified payload launched by Shuttle Atlantis crew.
Mar. 14 Intelsat 6 F-3 21A Titan 3	Objective: To launch satellite into geosynchronous orbit. Spacecraft: Launch configuration 17.5 ft (5.3 m) long, with diameter of 12 ft (3.6 m). Five antennas and 2 cylindrical solar arrays. In orbit, telescoping spacecraft is nearly 39 ft (11.7 m) long. Weight at launch, including apogee kick stage: 27,425 lbs (12,466 kg).	338 166 89.3 28.6	Launch vehicle malfunctioned, stranding vehicle in useless orbit. Intelsat may request NASA to use Shuttle in rescue attempt.
Mar. 26 USA-54 25A Delta III	Objective: Development of space flight techniques and technology. Spacecraft: Not announced.	20,207 20,059 716.0 55.1	Seventh navigation satellite in series. Global Positioning Satellite (GPS).
Apr. 5 PEGSAT 28A Pegasus	Objective: To launch satellite and test new launch delivery system. Spacecraft: Two barium canisters, 50 inches (127 cm) high and 42 inches (106 cm) wide. Weight: 272 lbs.	682 500 96.4 94.1	NASA-built small payload, to obtain environmental measurements of Pegasus launch vehicle for DARPA, deploy Navy satellite, and conduct two NASA scientific chemical-release experiments over north-central Canada.

Launch Date (GMT), Spacecraft Name, COSPAR Designation, Launch Vehicle	Mission Objectives, Spacecraft Data	Apogee and Perigee (km), Period (min.) Inclination to Equator (°)	Remarks
Apr. 5 USA-55 28B	Objective: To launch small experimental Navy satellite. Spacecraft: Polygonal sphere. Weight: 150 lbs.	673 498 96.3 94.1	Navy-designed Small Experimental Communications Satellite (SECS) deployed from Pegsat.
Apr. 11 USA-56 31A Atlas E	Objective: Development of space flight techniques and technology. Spacecraft: Not announced.	Not announced	First of triple payload launch.
Apr. 11 USA-57 31B	Objective: Development of space flight techniques and technology. Spacecraft: Not announced.	Not announced	Second of triple payload launch.
Apr. 11 USA-58 31C	Objective: Development of space flight techniques and technology. Spacecraft: Not announced.	Not announced	Third of triple payload launch.
Apr. 13 PALAPA-B2R 34A Delta	Objective: To launch spacecraft into geosynchronous orbit. Spacecraft: Not announced.	37,785 35,717 1,485.1 0.4	Indonesian communications satellite launched by United States.
Apr. 24 Space Shuttle Discovery (STS-31) 37A	Objective: To launch Hubble Space Telescope (HST) and conduct scientific and medical experiments while in weightless environment. Spacecraft: Shuttle orbiter carrying satellite as well as experiments: Ascent Particle Monitor (APM), IMAX camera, Investigation into Polymer Membrane Processing (IPMP),Protein Crystal Growth (PCG), Radiation Monitoring Experiment (RME), Ion Arc (student experiment), and Air Force Maui Optical Site (AMOS) Calibration Test. Weight of orbiter and cargo at main engine cutoff: 259,229 lbs.	619 613 96.8 28.4	Thirty-fifth flight of Space Transportation System (STS). Piloted by Loren J. Shriver and Charles F. Bolden, Jr. Mission specialists Steven A. Hawley, Bruce McCandless II, and Kathryn D. Sullivan. Discovery launched from KSC at 8:33 a.m. EDT. Shuttlelanded at Edwards AFB, CA, at 6:49 a.m. PDT on Apr. 29. Mission duration: 5 days, 1 hr, 16 min.

Launch Date (GMT), Spacecraft Name, COSPAR Designation, Launch Vehicle	Mission Objectives, Spacecraft Data	Apogee and Perigee (km), Period (min.) Inclination to Equator (º)	Remarks
Apr. 25 Hubble Space Telescope 37B	Objective: To perform variety of astronomical observations as long-term (15-year) international observatory with many different scientific goals and observational modes. Spacecraft: Cylindrical body made up of three major elements: (1) support systems module; (2) optical telescope assembly; and (3) scientific instruments. Electrical power provided by two solar arrays containing 48,000 solar cells. Optical telescope assembly contains two mirrors: 94-inch primary mirror located near center of Hubble Space Telescope (HST) and 13-inch secondary mirror located 16 ft in front of primary mirror. The two mirrors must remain in precise alignment for the images to be in focus. HST scientific instruments are the Widefield/Planetary Camera, Goddard High Resolution Spectrograph, Faint Object Spectrograph, and High Speed Photometer. Weight at launch: approximately 25,000 lbs (11,355.4 kg).	620 611 96.8 28.4	Deployed by Shuttle orbiter Discovery. First images taken by Widefield/Planetary Camera released May 20. Faint Object Camera (FOC) became first scientific instrument to complete its orbital verification activities, on Jun. 6. Spherical aberration in Hubble Space Telescope (HST) optical system announced publicly Jun. 27. Problem traced to spacing error in reflective null corrector, an optical reference device used in manufacture of primary mirror. Widefield/Planetary Camera was directed to study developing storm on Saturn; images expected to be 3 to 10 times better than those from Earth-based telescopes.
May 9 M-1 43A Scout	Objective: Development of spaceflight techniques and technology. Spacecraft: Not announced.	783 641 96.8 89.8	First of dual Lightsat launch.
May 9 M-2 43B	Objective: Development of spaceflight techniques and technology. Spacecraft: Not announced.	782 640 98.6 89.8	Second of dual Lightsat launch.

Launch Date (GMT), Spacecraft Name, COSPAR Designation, Launch Vehicle	Mission Objectives, Spacecraft Data	Apogee and Perigee (km), Period (min.) Inclination to Equator (º)	Remarks
Jun. 1 ROSAT 49A Delta II	Objective: To conduct an all-sky survey for six months, using imaging telescopes to measure positions of x-ray and extreme, ultraviolet (XUV) sources, while obtaining fluxes and special information. Spacecraft: Square-shaped central body containing scientific instruments, with three solar arrays. Launch configuration dimensions 4.5 m x 2.2 m x 2.4 m. Instrumentation includes most powerful x-ray telescope ever built, with secondary telescope a Widefield/Planetary Camera. Weight at launch: 2,424 kg.	588 567 96.1 52.9	Roentgen Satellite (ROSAT) international cooperative satellite successfully launched from Cape Canaveral Air Force Station, FL. Joint mission with Federal Republic of Germany and United Kingdom. Two-phase schedule: (1) six-month survey to be conducted by German scientists; and (2) 12-month pointed phase, where telescopes are pointed at preselected individual x-ray sources.
Jun. 8 USA-59 50A Titan 4	Objective: Development of space flight techniques and technology. Spacecraft: Not announced.	Not announced	First of four satellites launched by single booster.
Jun. 8 USA-60 50B	Objective: Development of space flight techniques and technology. Spacecraft: Not announced.	Not announced	Second of four satellites launched by single booster.
Jun. 8 USA-61 50C	Objective: Development of space flight techniques and technology. Spacecraft: Not announced.	Not announced	Third of four satellites launched by single booster.
Jun. 8 USA-62 50D	Objective: Development of space-flight techniques and technology. Spacecraft: Not announced.	Not announced	Fourth of four satellites launched by single booster.
Jun. 12 INSAT 1D 51A Delta	Objective: To launch communications satellite into geosynchronous orbit. Spacecraft: Same basic configuration as INSAT 1B launched by Shuttle in 1983.	35,974 35,767 1,440.0 0.2	Launched by United States for India to replace INSAT 1B. Completes INSAT 1 series.
Jun. 23 Intelsat 6 F-4 56A Titan 3	Objective: To launch satellite into geosynchronous orbit. Spacecraft: Same as Intelsat 6 launched Mar. 14.	35,791 35,785 1,436.2 0.0	Third in series of five Intelsat 6 satellites. Launched by Martin Marietta for Intelsat. Operational in early Sept.

Launch Date (GMT), Spacecraft Name, COSPAR Designation, Launch Vehicle	Mission Objectives, Spacecraft Data	Apogee and Perigee (km), Period (min.) Inclination to Equator (º)	Remarks
Jul. 25 CRRES 65A Atlas 1/Centaur 69	Objective: To launch satellite into highly elliptical geosynchronous transfer orbit to enable performance of active chemical release experiments in ionosphere and magnetosphere. Spacecraft: Octagonal satellite with two attached solar arrays and deployable magnetometer boom and 24 releasable canisters. Weight: 3,842 lbs.	33,612 335 591.9 18.2	Combined Release and Radiation Effects Satellite, joint NASA/Air Force payload, launched atop first commercial Atlas launch vehicle. First canister released over South Pacific Sept. 10.
Aug. 1 Soyuz TM-10 USSR 1990-067A			Crew consisted of Gennadiy Manakov and Gennadiy Strekalov. Docked with MIR space station. Crew returned Dec. 10 with Toyohiro Akiyama, Japanese astronaut. Flight time: 130 days, 20 hrs, 36 min.
Aug. 2 USA-63 68A Delta II	Objective: To place satellite into orbit to achieve Navy objectives. Spacecraft: Not announced.	20,665 19,931 722.7 54.7	Eighth navigation satellite. Global Positioning Satellite (GPS).
Aug. 18 BSB-R2 74A Delta	Objective: To launch satellite. Spacecraft: Not announced.	35,859 35,565 1,432.2 0.3	Launched for British Satellite Broadcasting.
Oct. 1 USA-64 88A Delta II	Objective: To place satellite into orbit to achieve Navy objectives. Spacecraft: Not announced.	20,378.0 19,984.0 717.9 55.0	Ninth in series of Block II operational NAVSTAR Global Positioning Satellites (GPS) launched by Air Force expendable launch vehicle. Operational system to be composed of 24 satellites in six orbital planes.

Launch Date (GMT), Spacecraft Name, COSPAR Designation, Launch Vehicle	Mission Objectives, Spacecraft Data	Apogee and Perigee (km), Period (min.) Inclination to Equator (º)	Remarks
Oct. 6 Space Shuttle Discovery (STS-41) 90A Oct. 6 Ulysses 90B	Objective: To launch Ulysses spacecraft toward rendezvous with the sun. To investigate interstellar space and the sun. Spacecraft: *Shuttle orbiter carrying spacecraft* with Inertial Upper Stage (IUS) booster and Payload Assist Module (PAM-5) with additional experiments: Chromosome and Plant Cell Division in Space Experiment (CHROMEX-2), Solid Surface Combustion Experiment (SSCE), Shuttle Solar Backscatter Ultra-violet (SSBUV) Instrument, Intelsat Solar Array Coupon (ISAC) experiment, Physiological Systems Experiment (PSE), Investigations into Polymer Membrane Processing (IPMP), *Voice Command System (VCS),* and Radiation Monitoring Equipment-III. Weight of experiments: 1,671 lbs.	303.0 280.0 90.2 28.4	Thirty-sixth flight of Space Transportation System (STS). Piloted by Richard N. Richards and Robert D. Cabana. Mission specialists Bruce E. Melnick, William M. Shepherd, and Thomas D. Akers. Discovery launched from KSC at 7:47 a.m. EDT. Shuttle landed at Edwards AFB, CA, at 9:57 a.m. EDT on Oct. 10. Mission duration: 4 days, 2 hrs, 10 min, 54 seconds.
	Objective: To investigate properties of solar wind, structure of sun/wind interface, heliospheric magnetic field, solar radio bursts and plasma waves, solar x-rays, solar and galactic cosmic rays, and interstellar interplanetary neutral gas and dust. Spacecraft: Spin-stabilized main bus, 10.5 x10.8x6.9 ft, 5.4 ft in diameter, parabolic high-gain antenna attached to main bus. After release from Shuttle payload bay, spacecraft will deploy 18.2 ft radial boom, 238 ft dipole wire boom, and 26.2 ft agial boom. Booms serve as antennas for radio wave-plasma experiment. Power source is radioisotope thermoelectric generator (RTG) attached to main bus. Weight, including Inertial Upper Stage (IUS) booster and Payload Assist Module (PAM-5): 44,024 lbs.	Heliocentric orbit.	Deployed by space Shuttle Discovery. Ulysses is a joint mission conducted by European Space Agency and NASA. Ulysses will encounter Jupiter in Feb. 1992, when spacecraft will receive gravity assisted velocity increase, enabling it to dive downward and away from ecliptic plane. Spacecraft will reach 70° south solar latitude in Jun. 1994.

Launch Date (GMT), Spacecraft Name, COSPAR Designation, Launch Vehicle	Mission Objectives, Spacecraft Data	Apogee and Perigee (km), Period (min.) Inclination to Equator (º)	Remarks
Nov. 13 USA-65 95A Titan 4	Objective: Development of space-flighttechniques and technology. Spacecraft: Not announced.	Not announced	Achieved orbit.
Nov. 15 Space Shuttle Atlantis (STS-38) 97A	Objective: To launch classified DoD payload. Spacecraft: Shuttle orbiter carrying classified DoD payload/cargo. Weight: Not announced.	221.0 215.0 88.6 28.4	Thirty-seventh flight of Space Transportation System. Piloted by Richard O. Covey and Frank L. Culberston, Jr. Mission specialists Charles "Sam" Gemar, Robert C. Springer, and Carl J. Meade. Atlantis launched from KSC at 6:48 p.m. EDT. Shuttle landed at KSC at 4:43 p.m. EST on Nov. 20. First KSC Shuttle landing since 1985. Mission duration: 4 days, 21 hrs, and 55 min.
Nov. 15 USA-67 97B	Objective: Development of space flight techniques and technology. Spacecraft: Not announced.	Not announced	Launched from Shuttle Atlantis.
Nov. 26 USA-66 103A Delta II	Objective: To place Navy satellite in orbit to achieve mission objectives. Spacecraft: Not announced.	20,279.0 19,935.0 714.8 54.8	Tenth spacecraft in series of operational NAVSTAR Global Positioning Satellites (GPS).
Dec. 1 USA-68 105A Atlas E	Objective: Development of space flight techniques and technology. Spacecraft: Not announced.	845.0 729.0 100.6 98.9	One in series of Defense Meteorological Satellites.

Launch Date (GMT), Spacecraft Name, COSPAR Designation, Launch Vehicle	Mission Objectives, Spacecraft Data	Apogee and Perigee (km), Period (min.) Inclination to Equator (º)	Remarks
Dec. 2 Soyuz TM-11 USSR 1990-107A			Crew consisted of Viktor Afanasyev and Musa Manarov. Docked with MIR space station. Toyohiro Akiyama returned Dec. 10 with previous MIR crew of Gennadiy Manakov and Gennadiy Strekalov. Flight time: 7 days, 21 hrs, 55 min.
Dec. 2 Space Shuttle Columbia (STS-35) 106A	Objective: To carry ASTRO-1 astrophysical observatory into orbit and return to Earth. Spacecraft: Shuttle orbiter Columbia carrying ASTRO-1 to cargo bay. ASTRO-1 to provide around-the-clock observations and measurements of ultraviolet radiation from celestial objects. Instruments include the ultraviolet astronomy observatory (ASTRO), Broad Band X-Ray Telescope (BBXRT), Hopkins Ultraviolet Telescope (HUT), Wisconsin Ultraviolet Photopolarimeter Experiment (WUPPE), and Ultraviolet Imaging Telescope (UIT). Weight: 27,454 lbs.	363.0 350.0 91.7 28.5	Thirty-eighth flight of Space Transportation System (STS). Piloted by Vance D. Brand and Guy S. Gardner. Mission specialists Jeffrey A. Hoffman, John M. "Mike" Lounge, and Robert A. R. Parker. Payload specialists Samuel T. Durrance and Ronald A. Parise. Columbia launched from KSC at 1:49 a.m. EST. Shuttle landed at Edwards AFB, CA, at 12:54 a.m. EST on Dec. 11. Mission duration: 8 days, 23 hrs, 5 min.

Appendix B

ABBREVIATIONS OF REFERENCES

Listed here are the abbreviations used for citing sources in the text. Not all the sources are listed, only those that are abbreviated.

Aero Space Rep	Aeronautics and Space Report of the President
AFP	Agence France Presse
AP	Associated Press new service
ARC Release	NASA Ames Research Center news release
AvWk	*Aviation Week & Space Technology* magazine
B Sun	*Baltimore Sun* newspaper
Bus Wk	*Business Week* magazine
C Trib	*Chicago Tribune* newspaper
CSM	*Christian Science Monitor* newspaper
D News	*Detroit News* newspaper
Def News	*Defense News*
FBIS-Chi	Foreign Broadcast Information Service, Chinese number
FBIS-Eas	Foreign Broadcast Information Service, Eastern Europe number
FBIS-Lat	Foreign Broadcast Information Service, Latin American number
FBIS-Nes	Foreign Broadcast Information Service, Near East number
FBIS-Sov	Foreign Broadcast Information Service, Soviet number
FBIS-Weu	Foreign Broadcast Information Service, Western Europe number
Fla Today	*Florida Today* newspaper
GSFC Release	NASA Goddard Space Flight Center news release
H Chron	*Houston Chronicle* newspaper
H Post	*Houston Post* newspaper
Htsvl Tms	*Huntsville Times* newspaper
JPL Voyager Status Report	NASA Jet Propulsion Laboratory, Voyager Status Report
JSC Release	NASA Lyndon B. Johnson Space Center news release
LA Herald	*Los Angeles Herald-Examiner* newspaper
LA Star News	*Los Angeles Star News* newsaper
LA Times	*Los Angeles Times* newspaper

LRC Release . . . . .Langley Research Center news release

M News . . . .*Miami News* newspaper

MSFC Release .NASA George C. Marshall Space Flight Center
news release

NASA anno . . . . .NASA announcement

NASA DAR . . . . .NASA Daily Activities Report

NASA MOR . . . . .NASA Headquarters Mission Operations Report,
preliminary launch and postlaunch report series

NASA PFOR . . . . .NASA Post-Flight Operations Report

NASA Release . . . . .NASA Headquarters news release

NASA spl anno . . . . .NASA special announcement

Nat Sp Trans Sys . . . . .National Space Transportation Critical Items List

NY Times . . . .*New York Times* newspaper

O Sen Star . . . .*Orlando Sentinel Star* newspaper

P Inq . . . .*Philadelphia Inquirer* newspaper

Reuters . . . . .Reuters Press Agency

Science . . . .*Science* magazine

SF Chron . . . .*San Francisco Chronicle* newspaper

SP News . . . .*Space Propulsion* newsletter

SSR . . . . .NASA Satellite Situation Report

Tor Star . . . .*Toronto Star* newspaper

UPI . . . . .United Press International news service

USA Today . . . .*USA Today* newspaper

W Post . . . .*Washington Post* newspaper

W Times . . . .*Washington Times* newspaper

WH Release . . . . .White House news release

WSJ . . . .*Wall Street Journal* newspaper

Index

About the Compilers

Ihor Y. Gawdiak, a senior research analyst at the Library of Congress Federal Research Division, has done extensive research and written numerous government studies on Soviet and East European political and military topics. Since the demise of the Soviet Union, he has continued to research and write primarily on political and military developments in the Newly Independent States. He is the author of the "Nationalities and Religions" chapter in the *Soviet Union: A Country Study* and editor of *Czechoslovakia: A Country Study*, both published by the Department of the Army as part of its Country Studies/Area Handbook Program. In addition, he is the author of the *NASA Historical Data Book, Volume IV, NASA Resources 1969–1978*. He is also the editor of *Ukraine: A Country Study* and *The Czech Republic: A Country Study*, both of which are are scheduled for publication at a later date. Mr. Gawdiak received his B.A. and M.A. degrees in International Relations from Clark University in Worcester, Massachusetts. He has completed all requirements and is working on his dissertation in the doctoral program in Russian History at the University of Maryland.

Ramón J. Miró was a Latin American research analyst at the Library of Congress Federal Research division from 1992 to 1996. He has a B.A. in Political Science from the University of Florida and an M.A. in Latin American Studies from Georgetown University. He resigned from the Library of Congress in 1996 to accept a fellowship for a doctoral program in Latin American Studies at Georgetown.

Sam Stueland has been a research analyst at the Library of Congress Federal Research Division since 1992 where he has worked on a number of different projects. Prior to joining the Library of Congress, Mr. Stueland was a researcher at the Smithsonian Institute National Museum of American History, Department of History of Science and Technology. He graduated Summa Cum Laude in American History from Gallaudet University in 1991. He has published articles in *"Technology and Culture"* and *"Railroad History."*

The NASA History Series

Reference Works, NASA SP-4000:

Grimwood, James M. *Project Mercury: A Chronology.* (NASA SP-4001, 1963).

Grimwood, James M., and Hacker, Barton C., with Vorzimmer, Peter J. *Project Gemini Technology and Operations: A Chronology.* (NASA SP-4002, 1969).

Link, Mae Mills. *Space Medicine in Project Mercury.* (NASA SP-4003, 1965).

Astronautics and Aeronautics, 1963: Chronology of Science, Technology, and Policy. (NASA SP-4004, 1964).

Astronautics and Aeronautics, 1964: Chronology of Science, Technology, and Policy. (NASA SP-4005, 1965).

Astronautics and Aeronautics, 1965: Chronology of Science, Technology, and Policy. (NASA SP-4006, 1966).

Astronautics and Aeronautics, 1966: Chronology of Science, Technology, and Policy. (NASA SP-4007, 1967).

Astronautics and Aeronautics, 1967: Chronology of Science, Technology, and Policy. (NASA SP-4008, 1968).

Ertel, Ivan D., and Morse, Mary Louise. *The Apollo Spacecraft: A Chronology, Volume I, Through November 7, 1962.* (NASA SP-4009, 1969).

Morse, Mary Louise, and Bays, Jean Kernahan. *The Apollo Spacecraft: A Chronology, Volume II, November 8, 1962-September 30, 1964.* (NASA SP-4009, 1973).

Brooks, Courtney G., and Ertel, Ivan D. *The Apollo Spacecraft: A Chronology, Volume III, October 1, 1964-January 20, 1966.* (NASA SP-4009, 1973).

Ertel, Ivan D., and Newkirk, Roland W., with Brooks, Courtney G. *The Apollo Spacecraft: A Chronology, Volume IV, January 21, 1966-July 13, 1974.* (NASA SP-4009, 1978).

Astronautics and Aeronautics, 1968: Chronology of Science, Technology, and Policy. (NASA SP-4010, 1969).

Newkirk, Roland W., and Ertel, Ivan D., with Brooks, Courtney G. *Skylab: A Chronology.* (NASA SP-4011, 1977).

Van Nimmen, Jane, and Bruno, Leonard C., with Rosholt, Robert L. *NASA Historical Data Book, Vol. I: NASA Resources, 1958-1968.* (NASA SP-4012, 1976, rep. ed. 1988).

Ezell, Linda Neuman. *NASA Historical Data Book, Vol II: Programs and Projects, 1958-1968.* (NASA SP-4012, 1988).

Ezell, Linda Neuman. *NASA Historical Data Book, Vol. III: Programs and Projects, 1969-1978.* (NASA SP-4012, 1988).

Astronautics and Aeronautics, 1969: Chronology of Science, Technology, and Policy. (NASA SP-4014, 1970).

Astronautics and Aeronautics, 1970: Chronology of Science, Technology, and Policy. (NASA SP-4015, 1972).

Astronautics and Aeronautics, 1971: Chronology of Science, Technology, and Policy. (NASA SP-4016, 1972).

Astronautics and Aeronautics, 1972: Chronology of Science, Technology, and Policy. (NASA SP-4017, 1974).

Astronautics and Aeronautics, 1973: Chronology of Science, Technology, and Policy. (NASA SP-4018, 1975).

Astronautics and Aeronautics, 1974: Chronology of Science, Technology, and Policy. (NASA SP-4019, 1977).

Astronautics and Aeronautics, 1975: Chronology of Science, Technology, and Policy. (NASA SP-4020, 1979).

Astronautics and Aeronautics, 1976: Chronology of Science, Technology, and Policy. (NASA SP-4021, 1984).

Astronautics and Aeronautics, 1977: Chronology of Science, Technology, and Policy. (NASA SP-4022, 1986).

Astronautics and Aeronautics, 1978: Chronology of Science, Technology, and Policy. (NASA SP-4023, 1986).

Astronautics and Aeronautics, 1979-1984: Chronology of Science, Technology, and Policy. (NASA SP-4024, 1988).

Astronautics and Aeronautics, 1985: Chronology of Science, Technology, and Policy. (NASA SP-4025, 1990).

Gawdiak, Ihor Y. Compiler. *NASA Historical Data Book, Vol. IV: NASA Resources, 1969-1978.* (NASA SP-4012, 1994).

Noordung, Hermann. *The Problem of Space Travel: The Rocket Motor.* Ernst Stuhlinger, and J.D. Hunley, with Jennifer Garland. Editors. (NASA SP-4026, 1995).

Astronautics and Aeronautics, 1991-1995: Chronology of Science, Technology, and Policy. (NASA SP-4027, 1997).

Management Histories, NASA SP-4100:

Rosholt, Robert L. *An Administrative History of NASA, 1958-1963.* (NASA SP-4101, 1966).

Levine, Arnold S. *Managing NASA in the Apollo Era.* (NASA SP-4102, 1982).

Roland, Alex. *Model Research: The National Advisory Committee for Aeronautics, 1915-1958.* (NASA SP-4103, 1985).

Fries, Sylvia D. *NASA Engineers and the Age of Apollo* (NASA SP-4104, 1992).

Glennan, T. Keith. *The Birth of NASA: The Diary of T. Keith Glennan,* edited by J.D. Hunley. (NASA SP-4105, 1993).

Seamans, Robert C., Jr. *Aiming at Targets: The Autobiography of Robert C. Seamans, Jr.* (NASA SP-4106, 1996).

Project Histories, NASA SP-4200:

Swenson, Loyd S., Jr., Grimwood, James M., and Alexander, Charles C. *This New Ocean: A History of Project Mercury.* (NASA SP-4201, 1966).

Green, Constance McL., and Lomask, Milton. *Vanguard: A History.* (NASA SP-4202, 1970; rep. ed. Smithsonian Institution Press, 1971).

Hacker, Barton C., and Grimwood, James M. *On Shoulders of Titans: A History of Project Gemini.* (NASA SP-4203, 1977).

Benson, Charles D. and Faherty, William Barnaby. *Moonport: A History of Apollo Launch Facilities and Operations.* (NASA SP-4204, 1978).

Brooks, Courtney G., Grimwood, James M., and Swenson, Loyd S., Jr. *Chariots for Apollo: A History of Manned Lunar Spacecraft.* (NASA SP-4205, 1979).

Bilstein, Roger E. *Stages to Saturn: A Technological History of the Apollo/Saturn Launch Vehicles.* (NASA SP-4206, 1980; paperback reprint 1996).

Compton, W. David, and Benson, Charles D. *Living and Working in Space: A History of Skylab.* (NASA SP-4208, 1983).

Ezell, Edward Clinton, and Ezell, Linda Neuman. *The Partnership: A History of the Apollo-Soyuz Test Project.* (NASA SP-4209, 1978).

Hall, R. Cargill. *Lunar Impact: A History of Project Ranger.* (NASA SP-4210, 1977).

Newell, Homer E. *Beyond the Atmosphere: Early Years of Space Science.* (NASA SP-4211, 1980).

Ezell, Edward Clinton, and Ezell, Linda Neuman. *On Mars: Exploration of the Red Planet, 1958-1978.* (NASA SP-4212, 1984).

Pitts, John A. *The Human Factor: Biomedicine in the Manned Space Program to 1980.* (NASA SP-4213, 1985).

Compton, W. David. *Where No Man Has Gone Before: A History of Apollo Lunar Exploration Missions.* (NASA SP-4214, 1989).

Naugle, John E. *First Among Equals: The Selection of NASA Space Science Experiments* (NASA SP-4215, 1991).

Wallace, Lane E. *Airborne Trailblazer: Two Decades with NASA Langley's Boeing 737 Flying Laboratory.* (NASA SP-4216, 1994).

Butrica, Andrew J. Editor. *Beyond the Ionosphere: Fifty Years of Satellite Communication.* (NASA SP-4217, 1997).

Butrica, Andrews J. *To See the Unseen: A History of Planetary Radar Astronomy.* (NASA SP-4218, 1996).

Center Histories, NASA SP-4300:

Rosenthal, Alfred. *Venture into Space: Early Years of Goddard Space Flight Center.* (NASA SP-4301, 1985).

Hartman, Edwin, P. *Adventures in Research: A History of Ames Research Center, 1940-1965.* (NASA SP-4302, 1970).

Hallion, Richard P. *On the Frontier: Flight Research at Dryden, 1946-1981.* (NASA SP-4303, 1984).

Muenger, Elizabeth A. *Searching the Horizon: A History of Ames Research Center, 1940-1976.* (NASA SP-4304, 1985).

Hansen, James R. *Engineer in Charge: A History of the Langley Aeronautical Laboratory, 1917-1958.* (NASA SP-4305, 1987).

Dawson, Virginia P. *Engines and Innovation: Lewis Laboratory and American Propulsion Technology.* (NASA SP-4306, 1991).

Dethloff, Henry C. *"Suddenly Tomorrow Came...": A History of the Johnson Space Center, 1957-1990.* (NASA SP-4307, 1993).

Hansen, James R. *Spaceflight Revolution: NASA Langley Research Center from Sputnik to Apollo* (NASA SP-4308, 1995).

Wallace, Lane E. *Flights of Discovery: An Illustrated History of the Dryden Flight Research Center.* (NASA SP-4309, 1996).

General Histories, NASA SP-4400:

Corliss, William R. *NASA Sounding Rockets, 1958-1968: A Historical Summary.* (NASA SP-4401, 1971).

Wells, Helen T., Whiteley, Susan H., and Karegeannes, Carrie. *Origins of NASA Names.* (NASA SP-4402, 1976).

Anderson, Frank W., Jr., *Orders of Magnitude: A History of NACA and NASA, 1915-1980.* (NASA SP-4403, 1981).

Sloop, John L. *Liquid Hydrogen as a Propulsion Fuel, 1945-1959.* (NASA SP-4404, 1978).

Roland, Alex. Editor. *A Spacefaring People: Perspectives on Early Spaceflight.* (NASA SP-4405, 1985).

Bilstein, Roger E. *Orders of Magnitude: A History of the NACA and NASA, 1915-1990.* (NASA SP-4406, 1989).

Logsdon, John M. General Editor. With Lear, Linda J., Warren-Findley, Jannelle, Williamson, Ray A., and Day, Dwayne A. *Exploring the Unknown: Selected Documents in the History of the U.S. Civil Space Program, Volume I: Organizing for Exploration.* (NASA SP-4407, 1995).

Logsdon, John M. General Editor. With Day, Dwayne A., and Launius, Roger D. *Exploring the Unknown: Selected Documents in the History of the U.S. Civil Space Program, Volume II: External Relationships.* (NASA SP-4407, 1996).